Elements of Modern Physics

Editors

Dr. S.K. Shrivastava

- **Dr. K. Chaturvedi**
- **Dr. S.K. Verma**
- **Dr. Man Singh**
- **Ashish Nautiyal**

SHREE PUBLISHERS & DISTRIBUTORS

NEW DELHI-110 002

Edition : 2012

Published by :
SHREE PUBLISHERS & DISTRIBUTORS
22/4735 Prakash Deep Building,
Ansari Road, Darya Ganj,
New Delhi–110 002

ISBN : 978–81–8329–482–9

Printed by :
D.G. Printers
Delhi–110 094

PREFACE

What is the necessity to edit these types of books? As reflected in the structure of the book itself, there are mainly two reasons. First, there are still a lot of new fundamental aspects to be learned about Acoustics, Condensed Matter Physics and Nuclear Physics and second, the potential of future discoveries remains too. The motivation to publish this book springs from a realization that physical sciences by their unique nature, pose special problems, challenges and opportunities in various areas of modern physics and this is an effort to present some selected material in this volume from the entire subject.

This book covers 20 Chapters related to Acoustics, Crystal Growth and Nuclear Physics and wide range of their applications. The first part of this book is dealt with some fundamental theories and applications of Acoustics. The second part is related to Solid State Physics and some mathematical models for electronic structures. Third and final part deals with Nuclear Physics where some topics of current interest, like neutrino oscillations, double beta decay and super heavy elements are discussed.

It is our wish that this volume will be of great help to students and to researchers interested in learning Ultrasonics, Solid State Physics and Nuclear Physics and its applications. We hope also that it can serve as a reference to workers in the field of Physical Sciences. We hope the book can contribute in a small way towards the rapid progress of the Physical Sciences. This book has been benefited from the work of many people, whose assistance We gratefully acknowledge. We like to take this opportunity to thank all researchers and investigators who are working significantly in the field of physical sciences. Finally, we are highly indebted our beloved parents whose love and affection has always been inspiring us to go ahead and ahead.

Editors

CONTRIBUTORS

Amit Kumar : Department of Physics, CSE, Ambabai, Jhansi, India.

Anchala : Department of Physics, Bundelkhand University, Jhansi-284128 UP India.

Anjita Srivastava : Department of Physics, Bundelkhand University, Jhansi-284128 UP India.

Arjun Mishra : Department of Physics, Bundelkhand University, Jhansi-284128 UP India.

Ashish Nautiyal : Department of Physics Bundelkhand University, Jhansi-284128 UP India.

Bhoopendra Singh : Department of Physics, Bundelkhand University, Jhansi-284128 UP India.

Jai Prakash : Department of Physics, Bundelkhand University, Jhansi-284128 UP India.

K. Chaturvedi : Department of Physics, Bundelkhand University, Jhansi-284128 UP India.

Kailash : Department of Physics, BNPG College, Rath-210431 UP India.

M. Tripathi : Department of Physics, Bundelkhand University, Jhansi-284128 UP India.

Man Singh : Department of Physics, Bundelkhand University, Jhansi-284128 UP India.

Omji Bajpai : Department of Physics, Bundelkhand University, Jhansi-284128 UP India.

Pushpraj Singh : Department of Physics Bundelkhand University, Jhansi-284128 UP India.

Rahul Gupta : Department of Physics, Bundelkhand University, Jhansi-284128 UP India.

Rajendra Kumar : Department of Physics, Bundelkhand University, Jhansi-284128 UP India.

S. K. Shrivastava : Department of Physics, Bundelkhand University, Jhansi-284128 UP India.

S.K. Verma : Department of Physics, Bundelkhand University, Jhansi-284128 UP India.

Sachin Shrivastava : Department of Physics Bundelkhand University, Jhansi-284128 UP India.

Shweta Goswami : Department of Physics, Bundelkhand University, Jhansi-284128 UP India.

Varun Yadav : Department of Physics, Bundelkhand University, Jhansi-284128 UP India.

Vidya : Department of Physics, Bundelkhand University, Jhansi-284128 UP India.

Contents

1

INTRODUCTION TO ULTRASONICS

S.K. Shrivastava, Kailash, Sachin Shrivastava and Anjita Shrivastavas

Introduction

Ultrasonics is the study and application of high-frequency sound waves, usually in excess of 20KHz (20,000 cycles per second). Modern ultrasonic generators can produce frequencies of as high as several gigahertz (several billion cycles per second) by transforming alternating electric currents into mechanical oscillations, and scientists have produced ultrasound with frequencies up to about10GHz (ten billion vibrations per second). There may be an upper limit to the frequency of usable ultrasound, but it is not yet known.

Higher frequencies have shorter wavelengths, which allows them to reflect from objects more readily and to provide better information about those objects. However, extremely high frequencies are difficult to generate and to measure. Detection and measurement of ultrasonic waves is accomplished mainly through the use of piezoelectric receivers or by optical means. The latter is possible because ultrasonic waves are rendered visible by the diffraction of light. Ultrasound is far above the range of human hearing, which is only about 20Hz to 18KHz. However, some mammals can hear well above this. For example, bats and whales use echo location that can reach frequencies in excess of 100KHz.

"Ultrasonics" should not be confused with the term "supersonics," which was formerly applied to this field.

Supersonics now refers to the study of phenomena arising when the velocity of a solid body exceeds the speed of sound.

Brief History

The roots of ultrasonic technology can be traced back to research on the piezoelectric effect conducted by Pierre Curie around 1880. He found that asymmetrical crystals such as quartz and Rochelle salt (potassium sodium tartrate) generate an electric charge when mechanical pressure is applied. Conversely, mechanical vibrations are obtained by applying electrical oscillations to the same crystals. One of the first applications for ultrasonics was sonar (an acronym for sound navigation ranging). It was employed on a large scale by the U.S. Navy during World War II to detect enemy submarines. Sonar operates by bouncing a series of high frequency, concentrated sound wave beams off a target and then recording the echo. Because the speed of sound in water is known, it is an easy matter to calculate the distance of the target. Prior to World War II researchers were inspired by sonar to develop analogous techniques for medical diagnosis. For example, the use of ultrasonic waves in detecting metal objects was discussed beginning in 1929. In 1931 a patent was obtained for using ultrasonic waves to detect flaws in solids.

Japan played an important role in the field of ultrasonics from an early date. For example, soon after the end of the war, researchers there began to explore the medical diagnostic capabilities of ultrasound. Japan was also the first country to apply Doppler ultrasound, which detects internal moving objects such as blood flowing through the heart.

In the 1950s researchers in the United States and Europe became increasingly aware of the progress that had been made in Japan, and they began work on additional medical applications.

The first ultrasonic instruments displayed their results with blips on an oscilloscope screen. That was followed by the use of two dimensional, gray scale imaging. High resolution, color, computer-enhanced images are now common, Ultrasonics technology is now employed in a wide range of applications in research, industry and medicine.

Cleaning

Perhaps the most common type of applications for ultrasonics is cleaning. This includes the removal of grease, dirt, rust and paint from metal, ceramic, glass and crystal surfaces of parts used in the electronic, automotive, aircraft, and precision instruments industries.

This cleaning is accomplished through the use of the cavitation effect. Cavitation is the rapid formation and collapse of tiny, gas and vapor filled bubbles or cavities in a solution that is irradiated with ultrasound. The repeated collapsing of these bubbles produces tiny shock waves that scrub the contaminants off of the surfaces of the parts. A variety of cleaning solutions can be used, including water, detergents and organic solvents.

Ultrasonic cleaning can be highly efficient for applications in which extreme cleanliness is required. It is also well suited for cleaning parts with very complex shapes.

Examples of specific applications are optical glass for lenses, quartz crystals, small ball bearings and dental bridges.

Flow Metering

Ultrasonic metering of flowing liquids is based on the Doppler effect. This type of metering has the advantages that it has no effect on the flow and can be used to monitor closed systems, such as a coolant in a nuclear power plant or the flow of blood to the human heart.

Non-Destructive Testing

Nondestructive testing has been practiced for many decades, with initial rapid developments in instrumentation spurred by the technological advances that occurred during World War II and the subsequent defense effort. Among the techniques that have been developed are eddy currents, x-rays, dye penetrants, magnetic particles and ultrasonics.

Ultrasonics is particularly attractive for non-destructive testing because it can be used with most types of materials, and it can be used to investigate both their surfaces and their interiors. The attenuation of ultrasonic waves is very low in solids and liquids, thus allowing solids more than 20 feet in thickness to be penetrated by both continuous and pulsed waves. Ultrasonic testing uses sound waves to detect imperfections in material and to measure material properties. The most commonly used ultrasonic testing technique is pulse-echo, wherein sound is introduced into a test object and reflections (echoes) returned to a receiver from internal imperfections or from the part's geometrical surfaces are analyzed. Defects and other internal irregularities result in changes in the echo pattern from the waves. The shadow method is used to inspect large castings and forgings, The echo reflection method is used primarily for the inspection of welds and castings. The primary purpose of ultrasonic testing was initially the detection of defects so that defective components could be removed from service. However, in the early 1970's the ability to detect small flaws led to the unsatisfactory situation that more and more parts had to be rejected, even though the probability of failure had not changed.

However, the discipline of fracture mechanics emerged, which enabled one to predict whether a crack of a given size would fail under a particular load if a material property, fracture toughness, were known. Other guidelines were developed to predict the rate of growth of cracks under cyclic loading (fatigue). This made it practical to accept structures containing defects if the sizes of those defects were known. This formed the basis for new philosophy of "fail safe" or "damage tolerant" design. Components having known defects could continue in service as long as it could be established that those defects would not grow to a critical, failure producing size. Thus it became necessary to not only detect flaws but to also obtain quantitative information about flaw size in order to make predictions of remaining life. These concerns, which were felt particularly strongly in the defense and nuclear power industries. They led to the emergence of quantitative nondestructive evaluation (QNDE) as a new discipline. Ultrasonic testing of forgings, castings and other metal parts has become standard. Examples of applications are axles for vehicles and machine parts. A somewhat different type of example is railroad rails which are already in use. Specially designed vehicles containing ultrasound equipment regularly travel on major railroad lines scanning the rails underneath them for cracks and other defects which are usually invisible to the human eye but which could eventually lead to a derailment.

Machining

Another industrial application for ultrasonics technology is the machining of materials. Ultrasonic machining has the advantage over conventional, mechanical machining techniques that it is well suited for processing unusual or complex shapes because no rotary tool is required. This technique can be used for very hard and highly abrasive materials because the actual cutting is done by an abrasive material in a liquid carrier rather than a bit or blade which is subject to abrasion. Among the materials that can be so processed are soft steel, ceramics, glass and tungsten carbide.

Soldering and Welding

Ultrasound has also proved to be very useful for joining materials. It can be used for both soldering and welding. In the case off soldering, the cavitation produced by high intensity ultrasonic waves destroys the oxide layer on aluminum, thus permitting parts to be joined with tin soldering materials without the use of flux. In ultrasonic welding, pressure and heat generated by the intense vibratory action of the material to be welded and an ultrasonic welding head allows a thin sheet of metal to be joined to a much thicker section. Ultrasonic techniques can likewise be used to weld pieces of similar or dissimilar plastic to each other.

Electronics

Ultrasonics is intimately related to the electronics industry. One reason is, of course, because ultrasonic waves are generated, detected and interpreted by electronic devices.

Also, ultrasonics technology is used extensively for the testing, cleaning and soldering of electronic components.

In addition, SAW (surface acoustic wave) filters are a type of electronic component which operates at ultrasonic frequencies. They are important for a growing range of electronics applications, including cellular phones and high performance TV receivers.

Materials Science

Applications in materials science include the determination of such properties of solids as compressibility, specific heat ratios and elasticity. Ultrasound can be used to produce an "acoustic microscope," which is able to visualize detail down to the one micron level. Goals can range from the determination of fundamental microstructural characteristics such as grain size, porosity and texture (preferred grain orientation) to material properties related to such failure mechanisms as fatigue, creep, and fracture toughness, applications which are sometimes quite challenging due to the existence of competing effects.

Sonochemistry

Most applications of ultrasound use low power waves which pass through materials without affecting their physical or chemical structure.

However, very high intensity ultrasound can be used to cause chemical and physical changes in materials. This is accomplished by violent cavitation which results from the waves, creating stress and intensely heating a localized area. Among the chemical processes which can be produced are acceleration of chemical reactions, oxidation, hydrolysis, polymerization, depolymerization and the production of emulsions. The recent development of high intensity ultrasound generators suitable for large volume materials processing are making such sonochemistry more economical for commercial application.

Agriculture

Ultrasound has been used to measure the thickness of fat layers on pigs and cows as part of livestock management. It has also been used in improve the quality of homogenized milk. A related application is pest control, including killing insects.

Oceanography

In addition to the tracking of submarines, oceanographic applications include mapping the contours of the sea bottom, discovering sunken ships and searching for schools of fish.

MEdical Applications

One of the most rapidly advancing areas of application is medicine. Ultrasound is used for imaging the human body and as a means of heating tissues to treat various ailments. It is also used to sterilize surgical instruments. Generally, the higher frequencies are used for medical imaging. The lower frequencies, 1 MHz or less, have longer wavelengths and greater amplitude for a given input energy, thus producing greater disruption of the medium. Among the many important advances in recent years have been higher resolution, real-time monitoring and color images. Ultrasonic scanning has the big advantage over x-rays that there are apparently no adverse health effects. For this reason, it has come into widespread use for monitoring the condition of the fetus as it grows in the womb. The increasingly high precision of such monitoring has made it possible to detect defects even at the very early stages of pregnancy.

Ultrasonic scanning has also become extremely useful for obtaining information about the flow of blood through the heart and about the condition of the heart valves. Other important diagnostic applications are the detection of kidney stones, gallstones and tumors.

An example of medical treatment applications is brain surgery, for which a sharply focused, high intensity beam can destroy diseased tissue with high precision. Ultrasound has also been used in the therapeutic treatment of arthritis, bursitis, contusions, lumbago and neuroma. There is still considerable controversy about the mechanism of such therapy. However, there is little doubt that it can be effective. One theory is that the benefits arise from the heating and possibly a "micromassage" resulting from the ultrasound.

Generation of Ultrasound

Ultrasonic waves can be generated using mechanical, electromagnetic and thermal energy sources. They can be produced in gasses (including air), liquids and solids.

Magnetostrictive transducers use the inverse magnetostrictive effect to convert magnetic energy into ultrasonic energy. This is accomplished by applying a strong alternating magnetic field to certain metals, alloys and ferrites. Piezoelectric transducers employ the inverse piezoelectric effect using natural or synthetic single crystals (such as quartz) or ceramics (such as barium titanate) which have strong piezoelectric behavior. Ceramics have the advantage over crystals in that it is easy to shape them by casting, pressing and extruding. The piezoelectric effect was first studied by Pierre Curie around 1880. He found that asymmetrical crystals such as quartz and Rochelle salt (potassium sodium tartrate) generate an electric charge when mechanical pressure is applied. Conversely, mechanical vibrations are obtained by applying electrical oscillations.

The Future

There is widespread agreement among researchers and scientists that ultrasonics is still in its infancy. This is evidenced by the fact that there is a great deal which is still not known about the field and the continued rapid rate of progress on nearly all aspects of it. Among the keys to further progress will be advances in materials (particularly piezoelectric materials for the transducers), in electronics and in computers (for interpreting and enhancing the results). Improvements in performance will be accompanied by further reductions in cost and increased diversity in the applications, likely including the development of some completely new uses.

2

EARLY DEVELOPMENT OF ULTRASONICS

S. K. Shrivastava, Kailash,
Anchala and Sachin Shrivastava

Introduction

In 1822, Daniel Colladen, a Swiss physicist/engineer and **Charles-Francois Sturm,** a mathematician, used an underwater bell in an attempt to calculate the speed of sound in the waters of Lake Geneva, Switzerland. In his experiment an underwater bell was struck simultaneously with ignition of gunpowder. The flash from the ignition was observed 10 miles away and compared with the arrival of the sound from the bell underwater heard through a trumpet-like device in the water. In spite of these crude instruments, they managed to determine that the speed of sound under water was 1435 metres/second, a figure not too different from what is known today.

As ultrasonics in general follows the principles delineated in acoustics, its development, particularly in the early years, is to some extent embedded in the broad developments in acoustics. The study of acoustics probably had its beginning with the Greek philosopher Pythagorus (6th Century B.C.), whose experiments on the properties of vibrating strings were so popular that they led to a tuning system that bears his name (the Sonometer). Aristotle (4th century BC) assumed (correctly) that a sound wave resonates in air through motion of the air; a philosphy-based hypothesis more than one of experimental physics. Vitruvius (1st century BC), determined the correct mechanism for the movement of sound waves, and he contributed substantially to the acoustic design of theatres, because he was an architect. Boethius (6th century AD), the Roman philosopher, documented several ideas relating science to music, including a suggestion that the human perception of pitch is related to the physical property of frequency.

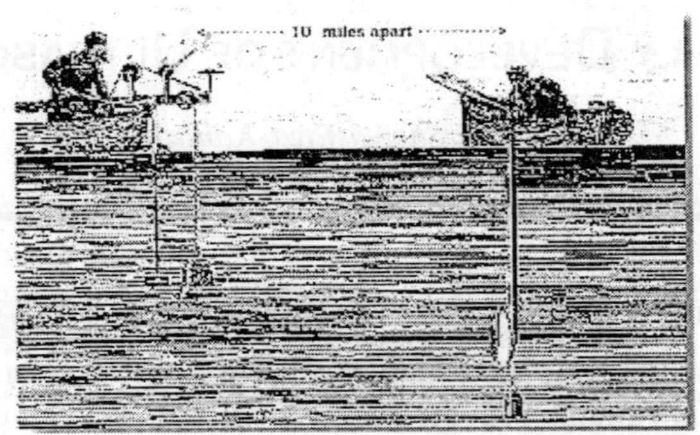

Galileo Galilei (1564-1642) is said to have started to modern studies of acoustics. He elevated the study of vibrations and the correlation between pitch and frequency of the sound source to scientific standards. His interest in sound was inspired in part by his father, who was a mathematician, musician, and composer. Following Galileo's foundation work, progress in acoustics came relatively quickly. The French mathematician Marin Mersenne studied the vibration of stretched strings; the results of these studies were summarized in the three Mersenne's laws. Mersenne's Harmonicorum Libri (1636) provided the basis for modern musical acoustics. Later in the century Robert Hooke, the English physicist, first produced a sound wave of known frequency, using a rotating cog wheel as a measuring device.

Further developed in the 19th century by the French physicist Félix Savart, and now commonly called Savart's disk, this device is often used today for demonstrations during physics lectures. In 1822, Swiss physicist Daniel Colladen used an underwater bell and successfully estimated the speed of sound in the waters of Lake Geneva. In the late 17th and early 18th centuries, detailed studies of the relationship between frequency and pitch and of waves in stretched strings were carried out by the French physicist Joseph Sauveur, who provided a legacy of acoustic terms used to this day and first suggested the name acoustics for the study of sound. In the early 1900's, lightships used a ranging system combination of an underwater gong and a foghorn on deck. The crew on approaching ships could hear both. The underwater sound was received by the use of a hydrophone on the hull. By timing the difference of the two sounds, they could determine their approximate distance from the lightship.

Investigations of high-frequency waves did not originate until the 19th century. The era of modern ultrasonics started about 1917, with Langevin's use of high-frequency acoustic waves and quartz resonators for submarine detection. Since that time, the field has grown enormously, with applications found in science, industry, medicine and other areas.

Underwater detection systems were developed for the purpose of underwater navigation by submarines in World war I and in particular after the Titanic sank in 1912. Alexander Belm in Vienna, described an underwater echo-sounding device in the same year. Within a week of the Titanic tragedy, Lewis Richardson filed a patent with the British patent office for echo ranging with airborne sound, following a month later with a patent application for the underwater equivalent. The first functioning echo ranger, however, was patented in the United States in 1914 by Canadian Reginald A. Fessenden, who worked for the Submarine Signal Company. Fessenden's device was an electric oscillator that emitted a low-frequency noise and then switched to a receiver to listen for echoes; it was able to detect an iceberg underwater from 2 miles away, although it could not precisely determine its direction.

Sergei Y. Sokolov at the V.I. Ulyanov (Lenin) Electrotechnical Institute proposed in 1928, and a few years later demonstrated a through-transmission technique for flaw detection in metals. He advanced his idea in the late 1920s, at a time when the required technology did not exist. He proposed that such technique could be used to detect irreguarities in solids such as metals. The resolution of the experimental devices which he fabricated was however poor and could not be used at a practical level.

Sokolov subsequently described a different and obviously the more important concept in ultrasonic applications. He demonstrated that sound waves could be used as a new form of microscope, basing on a reflective principle. Sokolov recognized that a 'microscope' using sound waves with a frequency of 3,000 megahertz (MHz) would have a resolution equal to that of the optical microscope. It was nevertheless not until the late 1930s that the technology for such devices was progressively developed, and the high frequencies required for Sokolov's microscope are found in microwave and ultrasonic systems used for radar and underwater navigation.

In the reflection technique, a pulsed sound wave is transmitted from one side of the sample, reflected off the far side, and returned to a receiver located at the starting point. Upon impinging on a flaw or crack in the material, the signal is reflected and its traveling time altered. The actual delay becomes a measure of the flaw's location; a map of the material can be generated to illustrate the location and geometry of the flaws. In the through-transmission method, the transmitter and receiver are located on opposite sides of the material; interruptions in the passage of sound waves are used to locate and measure flaws. Usually a water medium is employed in which transmitter, sample, and receiver are immersed.

The equipment suggested by Sokolov which could generate very short pulses necessary to measure the brief propagation time of their returning echoes was not available until the 1940s. Industrial use of ultrasonic testing apparently started spontaneously at around similar times in the United States and Great Britain. Such technology had also been in place in Germany and Japan, but developments had been curtailed because of the second world war.

The key-persons, Floyd Firestone, Donald Sproule and Adolf Trost had no knowledge of each other as they worked strictly in secret. Not even their patent-applications were published. Sproule

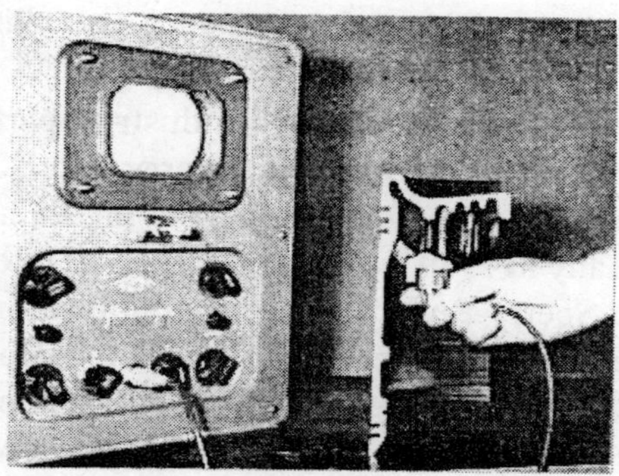

Fig. 1.

and Trost used transmission-technique with seperate transmitter- and receiver-probes. Trost invented the so-called "Trost-Tonge". The 2 probes were contacted on opposite sides of a plate, held in same axis by a mechanical device - the tonge - and coupled to both surfaces by continuously flowing water. Sproule placed the 2 probes on the same side of the workpiece. So he invented double-crystal probes. He used this combination also with variing distance from each other. Firestone was the first to realize the reflection-technique. He modified a radar instrument and developed a transmitter with short pulses and an amplifier with short dead-zone. Sproule eventually gave up the transmission method and filed a patent in 1952 entitled "the improvements in/ or relating to apparatus for flaw detection and velocity measurement by ultrasonic echo methods".

Firestone had made the following presentations and/or registered the following patents between 1940 and 1951:

- 1940 Flaw detecting device and measuring instrument
- 1943 Resonance inspection method; Surface and shear wave method and apparatus
- 1945 Supersonic reflectoscope, an instrument for inspecting the interior of solid parts by means of sound waves
- 1946 Refinements in supersonic reflectscope, Polarized sound
- 1948 Tricks with supersonic reflectscope

- 1951 Method and means for generating and utilizing vibrational wave in plate

Sproule and Firestone found industrial partners for their instruments: Kelvin-Hughes and Sperry Inc. Kelvin-Hughes produced their first commercial machine in 1952. Several years later, in Germany in 1949 two persons received information about the Firestone-Sperry-Reflectoscope by publications in technical papers: Josef Krautkrämer in Cologne and Karl Deutsch in Wuppertal. Both started developments - without knowledge of each other. Josef Krautkrämer and his brother Herbert were physicists, working in the field of oscilloscopes. They could develop ultrasonic instruments alone. Karl Deutsch, a mechanical engineer needed a partner for the electronics and found him with Hans-Werner Branscheid who had got some technical experience in radar-technique during the war. Within only one year both young and tiny companies could present their Ultrasonic testing-flaw-detectors, starting a competition still existing today.

Fig. 2. Early Pipe Testing with Krautkrämer Apparatus (1950s)

Later on more ultrasonic testing units came on the international markets: Siemens and Lehfeldt in Germany, Kretztechnik in Austria, Ultrasonique in France and Kelvin-Hughes in Britain. They all stopped their production before the 1970s. Kelvin-Hughes also stopped at the same time, Sperry was later renamed Automation Industries Inc.

Krautkrämer became world-wide market-leader in the early 60-ies and has kept this position until today. Besides Karl Deutsch new names came up: Nukem in Germany, Panametrics and Stavely (after Sonic and Harisonic) in USA, Sonatest and Sonomatic in Briatain, Gilardoni in Italy.

Research into ultrasonics and metal-flaw detection in Japan was considerably curtailed when World war II broke out in 1941, at which time the Americans and the Germans were both diligently researching into ultrasonics and the development of the Radar. The study into Radar techniques in Japan was also in the disadvantage. As the war ended in 1945, research into high-power electronics was prohibited in Japan for some time (up till 1948, when developments of non-military electronics resumed).

Japanese enterprises took on the research from the U.S. and England and soon developed its own flaw detectors in non-destructive testing. In about 1949, four Japanese companies started to manufacture their own flaw dectors. These were: the Mitsutbishi Electric Corporation, the Japan Radio Company (later became the Aloka Company), the Shimadsu Manufacturing Company and the Toyko Ultrasonic Industrial Company. Only Mitsubishi continued to expand in the field of non-destructive testing and the other companies moved on to other areas and in particular diagnostic medical ultrasound applications. The Japanese Society for Non-Destructive Inspection officially recognised the year 1952 as the first year non-destructive testing was implemented in Japan.

The use of Ultrasonics in the field of medicine had nonetheless started initially with it's applications in therapy rather than diagnosis, utilising it's heating and disruptive effects on animal tissues. The destructive ability of high intensity ultrasound had been recognised in the 1920s from the time of Langévin when he noted destruction of school of fishes in the sea and pain induced in the hand when placed in a water tank insonated with high intensity ultrasound; and from the seminal work in the late 1920s from Robert Wood and the legendary Alfred Loomis in New York.

In 1944, Lynn and Putnam successfully used ultrasound waves to destroy brain tissue in animals. William Fry and Russell Meyers performed craniotomies and used ultrasound to destroy parts of the basal ganglia in patients with Parkinsonism. Peter Lindstrom reported ablation of frontal lobe tissue in moribound patients to alleviate their pain from carcinomatosis. Ultrasonics was also extensively used in physical and rehabilitation medicine. Jerome Gersten reported in 1953 the use of ultrasound in the treatment of patients with rheumatic arthritis. Several groups such as the Peter Wells group in Bristol, England, the Mischele Arslan group in Padua, Italy and the Douglas Gordon group in London used ultrasonic energy in the treatment of Meniere's disease.

The major developments in Acoustics and Ultrasonics are summarized below :

1822 Colladen used underwater bell to calculate the speed of sound in waters of Lake Geneva.

1830 Savart developed large, toothed wheel to generate very high frequencies.

1842 Magnetostrictive effect discovered by Joule.

1845 Stokes investigated effect of viscosity on attenuation.

1860 Tyndall developed the sensitive flame to detect high frequency waves.

1866 Kundt used dust figures in a tube to measure sound velocity.

1876 Galton invented the ultrasonic whistle.

1877 Rayleigh's "Theory of Sound" laid foundation for modern acoustics.

1880 Curie brothers discovered the piezoelectric effect.

1890 Koenig, studying audibility limits, produced vibrations up to 90,000 Hz.

1903 Lebedev and coworkers developed complete ultrasonic system to study absorption of waves.

1912 Sinking of Titanic led to proposals on use of acoustic waves to detect icebergs.

1912 Richardson files first patent for an underwater echo ranging sonar.

1914 Fessenden built first working sonar system in the United States which could detect icebergs two miles away.

1915 Langevin originated modern science of ultrasonics through work on the"Hydrophone" for submarine detection.

1921 Cady discovered the quartz stabilized oscillator.

1922 Hartmann developed the air-jet ultrasonic generator.

1925 Pierce developed the ultrasonic interferometer.

1926 Boyle and Lehmann discovered the effect of bubbles and cavitation in liquids by ultrasound.

1927 Wood and Loomis described effects of intense ultrasound.

1928 Pierce developed the magnetostrictive transducer.

1928 Herzfeld and Rice developed molecular theory for dispersion and absorption of sound in gases.

1928 Sokolov proposed use of ultrasound for flaw detection.

1930 Debye and Sears and Lucas and Biquard discover diffraction of light by ultrasound.

1930 Harvey reported on the physical, chemical, and biological effects of ultrasound in macromolecules, microorganisms and cells.

1931 Mulhauser obtained a patent for using two ultrasonic transducers to detect flaws in solids.

1937 Sokolov invented an ultrasonic image tube.

1938 Pierce and Griffin detect the ultrasonic cries of bats.

1939 Pohlman investigated the therapeutic uses of ultrasonics.

1940 Firestone, in the United States and Sproule, in Britain, discovered ultrasonic pulse-echo metal-flaw detection.

1940 Sonar extensively developed and used to detect submarines.

1941 "Reflectoscopes" extensively developed for non-destructive metal testing.

1942 Dussik brothers made first attempt at medical imaging with ultrasound.

1944 Lynn and Putnam successfully used ultrasound waves to destroy brain tissue of animals.

1945 Newer piezoelectric ceramics such as barium titanate discovered.

1945 Start of the development of power ultrasonic processes.

1948 Start of extensive study of ultrasonic medical imaging in the United States and Japan.

1954 Jaffe discovered the new piezoelectric ceramics lead titanate-zirconate.

References

1. **Karl F. Graff**, "*A History of Ultrasonics, in Physical Acoustics*", Volume XV (Academic Press, New York, 1982).
2. **Peter B. Nagy**, "*An Introduction to Ultrasonics*", Department of Aerospace Engineering & Engineering Mechanics, University of Cincinnati, 2001.
3. **William D. O'Brien, Jr.**, "*Assessing the Risks for Modern Diagnostic Ultrasound Imaging*", Bioacoustics Research Laboratory, Department of Electrical and Computer Engineering, University of Illinois, 1997.

3

ULTRASONIC IMAGING

S. K. Shrivastava, Kailash,
Rajendra Kumar and Pushpraj Singh

Introduction

Sound is a physical phenomenon that transfers energy from one point to another. In this respect, it is similar to radiation. It differs from radiation, however, in that sound can pass only through matter and not through a vacuum as radiation can. This is because sound waves are actually vibrations passing through a material. If there is no material, nothing can vibrate and sound cannot exist. One of the most significant characteristics of sound is its frequency, which is the rate at which the sound source and the material vibrate. The basic unit for specifying frequency is the hertz, which is one vibration, or cycle, per second. Pitch is a term commonly used as a synonym for frequency of sound.

The human ear cannot hear or respond to all sound frequencies. The range of frequencies that can be heard by a normal young adult is from approximately 20 Hz to 20,000 Hz (20 kHz). Ultrasound has a frequency above this range. Frequencies in the range of 2 MHz (million cycles per second) to 20 MHz are used in diagnostic ultrasound. Ultrasound is used as a diagnostic tool because it can be focused into small, well-defined beams that can probe the human body and interact with the tissue structures to form images. The transducer is the component of the ultrasound imaging equipment that is placed in direct contact with the patient's body. It performs several functions as will be described in detail later. It's first function is to produce the ultrasound pulses when electrical pulses are applied to it. A short time later, when echo pulses return to the body surface they are picked up by the transducer and converted back into electrical pulses that are then processed by the system and formed into an image. When a beam of ultrasound pulses is passed into a body, several things happen. Most of the ultrasound energy is absorbed and the beam is attenuated. This is undesirable and does not contribute to the formation of an image like in x-ray imaging. Some of the pulses will be reflected by internal body structures and send echoes back to the surface where they are collected by the transducer and used to form the image. Therefore, the general ultrasound image is a display of structures or reflecting surfaces in the body that produce echoes as illustrated below.

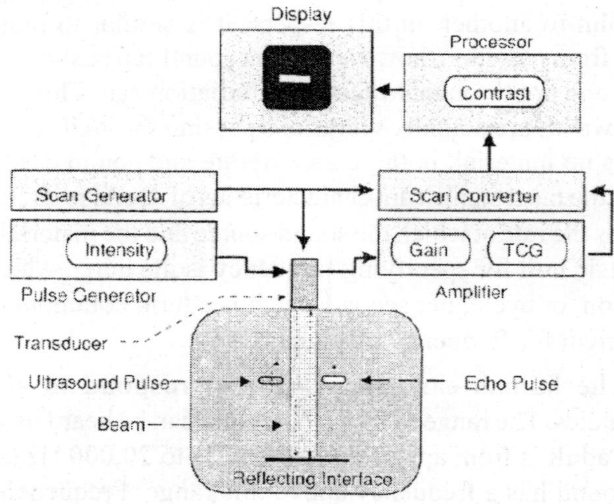

Fig. 1.

Another physical characteristic that can be imaged with ultrasound is motion, specifically the motion of flowing blood. This uses the Doppler principle and the images are usually displayed with different colors representing the different flow velocities and directions.

Physical Principle of Imaging

Echoes, which show up as bright or white spots in the image is produced by surfaces or boundaries between two different types of tissues. Most anatomical areas are composed of a "mixture" of different tissue types and many surfaces that produce the general gray and white background that we see in the image. Since there are no reflecting surfaces within a fluid, such as a cyst, it is dark in the image. Therefore, the general ultrasound image, sometimes called a "B mode" image, is a display of echo producing sites within the anatomical area. The basic functional components of an ultrasound imaging system are shown below.

We will now consider some of these functions in more detail and how they contribute to image formation. Modern ultrasound systems use digital computer electronics to control most of the functions in the imaging process. Therefore, the boxes in the illustration above represent functions performed by the computer and other electronic circuits and not individual physical components.

Imaging System

The ultrasound image is a display showing the location of reflecting structures or echo sites within the body. The location of a reflecting structure (interface) in the horizontal direction is determined by the position of the beam. In the depth direction, it is determined by the time required for the pulse to travel to the reflecting site and for the echo pulse to return.

Transducer

The transducer is the component of the ultrasound system that is placed in direct contact with the patient's body. It alternates between two major functions: (1) producing ultrasound pulses and (2) receiving or detecting the returning echoes. Within the transducer there are one or more piezoelectric elements. When an electrical pulse is applied to the piezoelectric element it vibrates and produces the ultrasound. Also, when the piezoelectric element is vibrated by the returning echo pulse it produces a pulse of electricity.

The transducer also focuses the beam of pulses to give it a specific size and shape at various depths within the body and also scans the beam over the anatomical area that is being imaged.

Pulse Generator

The pulse generator produces the electrical pulses that are applied to the transducer. For conventional ultrasound imaging the pulses are produced at a rate of approximately 1,000 pulses per second. This is the pulse rate (pulses per second) and not the frequency which is the number of cycles or vibrations per second within each pulse. The principal control associated with the pulse generator is the size of the electrical pulses that can be used to change the intensity and energy of the ultrasound beam.

Amplification

Amplification is used to increase the size of the electrical pulses coming from the transducer after an echo is received.. The amount of amplification is determined by the gain setting. The principal control associated with the amplifier is the time gain compensation (TGC), which allows the user to adjust the gain in relationship to the depth of echo sites within the body.

Scan Generator

The scan generator controls the scanning of the ultrasound beam over the body section being imaged. This is usually done by controlling the sequence in which the electrical pulses are applied to the piezoelectric elements within the transducer.

Scan Converter

Scan conversion is the function that converts from the format of the scanning ultrasound beam into a digital image matrix format for processing and display.

Image Processor

The digital image is processed to produce the desired characteristics for display. This includes giving it specific contrast characteristics and reformatting the image if necessary.

Display

The digital ultrasound images are viewed on the equipment display (monitor) and usually transferred to the physician display or work station.One component of the ultrasound imaging system that is not shown is the digital storage device that is used to store images for later viewing if that process is used.

The Ultrasound Pulse

The basic principles of ultrasound pulse production and transmission are illustrated below.

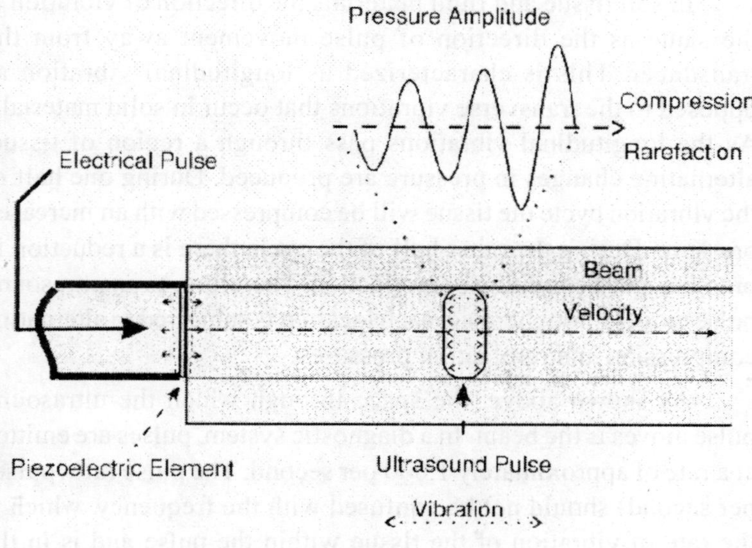

Fig. 2.

The Production of an Ultrasound Pulse

The source of sound is a vibrating object, the piezoelectric transducer element. Since the vibrating source is in contact with the tissue, it is caused to vibrate. The vibrations in the region of tissue next to the transducer are passed on to the adjacent tissue. This process continues, and the vibrations, or sound, is passed along from one region of tissue to another. The rate at which the tissue structures vibrate back and forth is the frequency of the sound. The rate at which the vibrations move through the tissue is the velocity of the sound.

The sound in most diagnostic ultrasound systems is emitted in pulses rather than a continuous stream of vibrations. At any instant, the vibrations are contained within a relatively small volume of the material. It is this volume of vibrating material that is referred to as the ultrasound pulse. As the vibrations are passed from one region of material to another, the ultrasound pulse, but not the material, moves away from the source.

In soft tissue and fluid materials the direction of vibration is the same as the direction of pulse movement away from the transducer. This is characterized as longitudinal vibration as opposed to the transverse vibrations that occur in solid materials. As the longitudinal vibrations pass through a region of tissue, alternating changes in pressure are produced. During one half of the vibration cycle the tissue will be compressed with an increased pressure. During the other half of the cycle there is a reduction in pressure and a condition of rarefaction. Therefore, as an ultrasound pulse moves through tissue, each location is subjected to alternating compression and rarefaction pressures.

As shown above, the space through which the ultrasound pulse moves is the beam. In a diagnostic system, pulses are emitted at a rate of approximately 1,000 per second. The pulse rate (pulses per second) should not be confused with the frequency, which is the rate of vibration of the tissue within the pulse and is in the range of 2-20 MHz.

Ultrasound Pulse Frequency

The frequency of ultrasound pulses must be carefully selected to provide a proper balance between image detail and depth of penetration. In general, high frequency pulses produce higher quality images but cannot penetrate very far into the body. These issues will be discussed in greater detail later.

The frequency of sound is determined by the source. For example, in a piano, the source of sound is a string that is caused to vibrate by striking it. Each string within the piano is adjusted, or tuned, to vibrate with a specific resonant frequency. In diagnostic ultrasound equipment, the sound is generated by the transducer. The major element within the transducer is a crystal designed to vibrate with the desired frequency. A special property of the crystal material is that it is piezoelectric. This means that the crystal will deform if electricity is applied to it. Therefore, if an electrical pulse is applied to the crystal it will have essentially the same effect as the striking of a piano string: the crystal will vibrate. If the transducer is activated by a single electrical pulse, the transducer will vibrate, or "ring," for a short period of time. This creates an ultrasound pulse as opposed to a continuous ultrasound wave. The ultrasound pulse travels into the tissue in contact with the transducer and moves away from the transducer surface, as shown in the above figure. A given transducer is often designed to vibrate with only one frequency, called its resonant frequency. Therefore, the only way to change ultrasound frequency is to change transducers. This is a factor that must be considered when selecting a transducer for a specific clinical procedure. Certain frequencies are more appropriate for certain types of examinations than others. Some transducers are capable of producing different frequencies. For these the ultrasound frequency is determined by the electrical pulses applied to the transducer.

Factors Related to Ultrasound Pulse Velocity

The significance of ultrasound velocity is that it is used to determine the depth location of structures in the body. The velocity with which sound travels through a medium is determined by the characteristics of the material and not characteristics of the sound. Most ultrasound systems are set up to determine distances using an assumed velocity of 1540 m/sec. This means that displayed depths will not be completely accurate in materials that produce other ultrasound velocities such as fat and fluid.

Wavelength

The distance sound travels during the period of one vibration is known as the wavelength, l. Although wavelength is not a unique property of a given ultrasound pulse, it is of some significance because it determines the size (length) of the ultrasound pulse. The illustration below shows both temporal and spatial (length) characteristics related to the wavelength. A typical ultrasound pulse consists of several wavelengths or vibration cycles. The number of cycles within a pulse is determined by the damping characteristics of the transducer. Damping is what keeps the transducer element from continuing to vibrate and produce a long pulse. The wavelength is determined by the velocity, **v**, and frequency, **f**, in this relationship: Wavelength (l) = v/f.

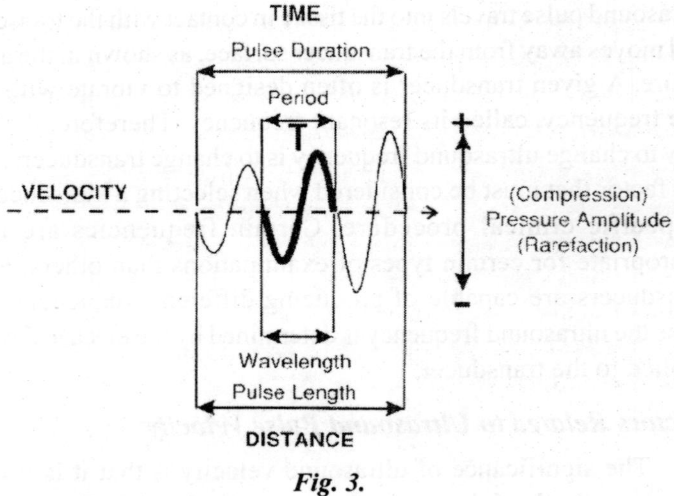

Fig. 3.

The Temporal and Length Characteristics of an Ultrasound Pulse

The period is the time required for one vibration cycle. It is the reciprocal of the frequency. Increasing the frequency decreases the period. In other words, wavelength is simply the ratio of velocity to frequency or the product of velocity and the period. This means that the wavelength of ultrasound is determined by the characteristics of both the transducer (frequency) and the material through which the sound is passing (velocity).

In ultrasound imaging the significance of wavelength is that short wavelengths are required to produce short pulses for good anatomical detail (in the depth direction) and this requires higher frequencies.

Ultrasound Pulse Amplitude, Intensity, and Energy

The pressure is related to the degree of tissue displacement caused by the vibration. The amplitude is related to the energy content, or "loudness," of the ultrasound pulse. The amplitude of the pulse as it leaves the transducer is generally determined by how hard the crystal is "struck" by the electrical pulse.

Most systems have a control on the pulse generator that changes the size of the electrical pulse and the ultrasound pulse amplitude. We designate this as the intensity control, although different names are used by various equipment manufacturers.

In diagnostic applications, it is usually necessary to know only the relative amplitude of ultrasound pulses. For example, it is necessary to know how much the amplitude, A, of a pulse decreases as it passes through a given thickness of tissue. The relative amplitude of two ultrasound pulses, or of one pulse after it has undergone an amplitude change, can be expressed by means of a ratio as follows:

$$\text{Relative amplitude (ratio)} = A_2/A_1$$

There are advantages in expressing relative pulse amplitude in terms of the logarithm of the amplitude ratio. When this is done the relative amplitude is specified in units of decibels (dB). The relative pulse amplitude, in decibels, is related to the actual amplitude ratio by

$$\text{Relative amplitude (dB)} = 20 \log A_2/A_1$$

When the amplitude ratio is greater than 1 (comparing a large pulse to a smaller one), the relative pulse amplitude has a positive decibel value; when the ratio is less than 1, the decibel value is negative. In other words, if the amplitude of a pulse is increased by some means, it will gain decibels, and if it is reduced, it will lose decibels.

The following illustration compares decibel values to pulse amplitude ratios and percent values. The first two pulses differ in amplitude by 1 dB. In comparing the second pulse to the first, this corresponds to an amplitude ratio of 0.89, or a reduction of approximately 11%. If the pulse is reduced in amplitude by another 11%, it will be 2 dB smaller than the original pulse. If the pulse is once again reduced in amplitude by 11 % (of 79%), it will have an amplitude ratio (with respect to the first pulse) of 0.71:1, or will be 3 dB smaller.

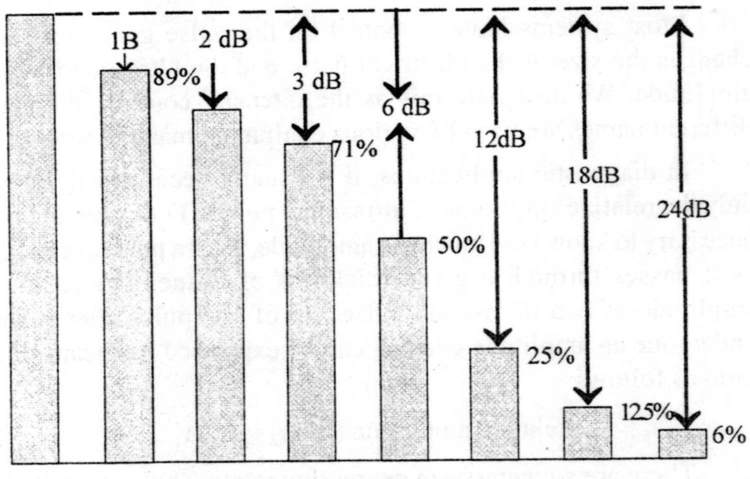

Fig. 4.

Intensity and Power

Power is the rate of energy transfer and is expressed in the units of watts. Intensity is the rate at which power passes through a specified area. It is the amount of power per unit area and is expressed in the units of watts per square centimeter. Intensity is the rate at which ultrasound energy is applied to a specific tissue location within the patient's body. It is the quantity that must be considered with respect to producing biological effects and safety. The intensity of most diagnostic ultrasound beams at the transducer surface is on the order of a few milliwatts per square centimeter.

Intensity is related to the pressure amplitude of the individual pulses and the pulse rate. Since the pulse rate is fixed in most systems, the intensity is determined by the pulse amplitude. The relative intensity of two pulses (I_1 and I_2) can be expressed in the units of decibels by:

$$\text{Relative Intensity} = 10 \log I_2/I_1$$

Note that when intensities are being considered, a factor of 10 appears in the equation rather than a factor of 20, which is used for relative amplitudes. This is because intensity is proportional to the square of the pressure amplitude, which introduces a factor of 2 in the logarithmic relationship. The intensity of an ultrasound beam is not constant with respect to time nor uniform with respect to spatial area, as shown in the following figure. This must be taken into consideration when describing intensity. It must be determined if it is the peak intensity or the average intensity that is being considered.

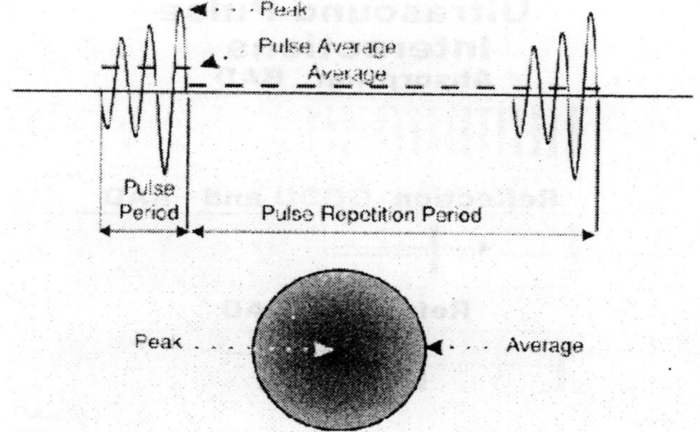

Fig. 4. Spatial Averaged Intensity

Temporal Characteristics

The figure above shows two sequential pulses. Two important time intervals are the pulse duration and the pulse repetition period. The ratio of the pulse duration to the pulse repetition period is the duty factor. The duty factor is the fraction of time that an ultrasound pulse is actually being produced. If the ultrasound is produced as a continuous wave (CW), the duty factor will have a value of 1. Intensity and power are proportional to the duty factor. Duty factors are relatively small, less than 0.01, for most pulsed imaging applications.

With respect to time there are three possible power (intensity) values. One is the peak power, which is associated with the time of maximum pressure. Another is the average power within a pulse. The lowest value is the average power over the pulse repetition period for an extended time. This is related to the duty factor.

Spatial Characteristics

The energy or intensity is generally not distributed uniformly over the area of an ultrasound pulse. It can be expressed either as the peak intensity, which is often in the center of the pulse, for as the average intensity over a designated area.

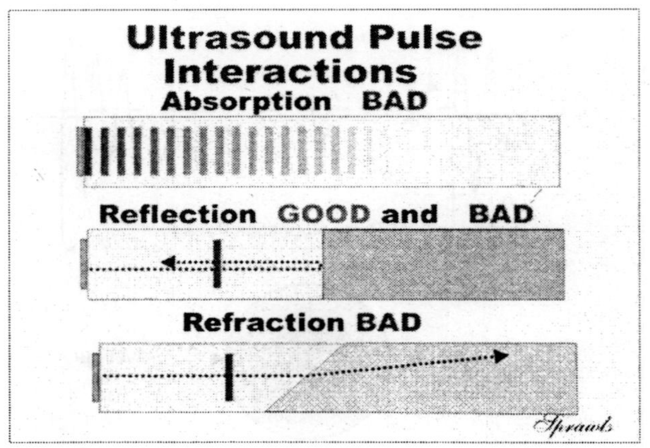

Fig. 5.

Temporal/Spatial Combinations

There is some significance associated with each of the intensity expressions. However, they are not all used to express the intensity with respect to potential biological effects. Thermal effects are most closely related to the spatial-peak and temporal-average intensity (I_{SPTA}). This expresses the maximum intensity delivered to any tissue averaged over the duration of the exposure. Thermal effects (increase in temperature) also depend on the duration of the exposure to the ultrasound. Mechanical effects such as cavitation are more closely related to the spatial-peak, pulse- average intensity (I_{SPPA}).

Absorption and Attenuation

As the ultrasound pulse moves through matter, it continuously loses energy. This is generally referred to as attenuation. Several factors contribute to this reduction in energy. One of the most significant is the absorption of the ultrasound energy by the material and its conversion into heat. Ultrasound pulses lose energy continuously as they move through matter. This is unlike x-ray photons, which lose energy in "one-shot" photoelectric or Compton interactions. Scattering and refraction interactions also remove some of the energy from the pulse and contribute to its overall attenuation, but absorption is the most significant.

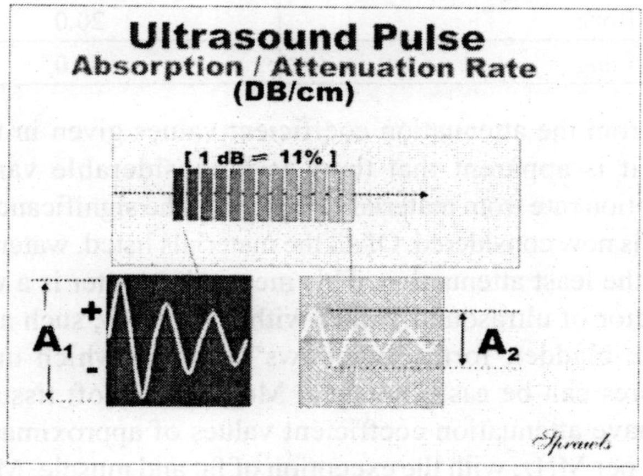

Fig. 6.

The rate at which an ultrasound pulse is absorbed generally depends on two factors: (1) the material through which it is passing, and (2) the frequency of the ultrasound. The attenuation (absorption) rate is specified in terms of an attenuation coefficient in the units of decibels per centimeter. Since the attenuation in tissue increases with frequency, it is necessary to specify the frequency when an attenuation rate is given. The attenuation through a thickness of material, x, is given by:

$$\text{Attenuation (dB)} = (a)\,(f)\,(x)$$

where a is the attenuation coefficient (in decibels per centimeter at 1 MHz), and f is the ultrasound frequency, in megahertz.

Approximate values of the attenuation coefficient for various materials of interest are given in the following table.

Approximate Attenuation Coefficient Values for Various Materials

Material	**Coefficient (dB/cm MHz)**
Water	0.002
Fat	0.66
Soft tissue (average)	0.9
Muscle (average)	2.0
Air	12.0
Bone	20.0
Lung	40.0

From the attenuation coefficient values given in the above table, it is apparent that there is a considerable variation in attenuation rate from material to material. The significance of these values is now considered. Of all the materials listed, water produces by far the least attenuation. This means that water is a very good conductor of ultrasound. Water within the body, such as in cysts and the bladder, forms "windows" through which underlying structures can be easily imaged. Most of the soft tissues of the body have attenuation coefficient values of approximately 1 dB per cm per MHz, with the exception of fat and muscle. Muscle has a range of values that depends on the direction of the ultrasound with respect to the muscle fibers. Lung has a much higher attenuation rate than either air or soft tissue. This is because the small pockets of air in the alveoli are very effective in scattering ultrasound energy. Because of this, the normal lung structure is extremely difficult to penetrate with ultrasound. Compared to the soft tissues of the body, bone has a relatively high attenuation rate. Bone, in effect, shields some parts of the body against easy access by ultrasound.

The following illustration shows the decrease in pulse amplitude as ultrasound passes through various materials found in the human body.

Reflection

The reflection of ultrasound pulses by structures within the body is the interaction that creates the ultrasound image. The reflection of an ultrasound pulse occurs at the interface, or boundary, between two dissimilar materials, as shown in the following figure. In order to form a reflection interface, the two materials must differ in terms of a physical characteristic known as acoustic impedance **Z**. Although the traditional symbol for

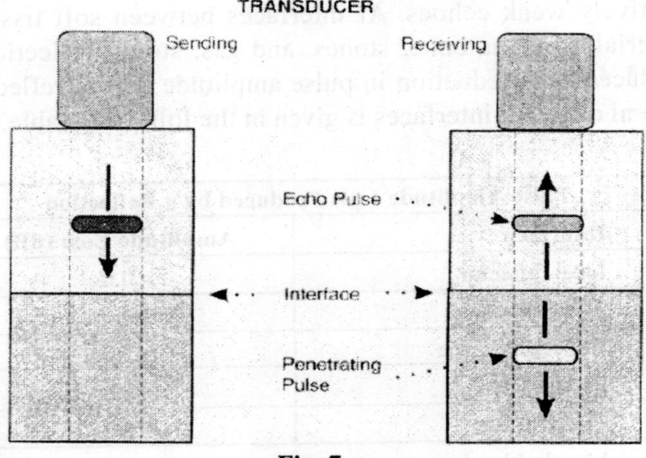

Fig. 7.

impedance, **Z**, is the same symbol used for atomic number, the two quantities are in no way related. Acoustic impedance is a characteristic of a material related to its density and elastic properties. Since the velocity is related to the same material characteristics, a relationship exists between tissue impedance and ultrasound velocity. The relationship is such that the impedance, **Z**, is the product of the velocity, **v**, and the material density, **Y**, which can be written as

$$\text{Impedance} = (Y)(v).$$

At most interfaces within the body, only a portion of the ultrasound pulse is reflected. The pulse is divided into two pulses, and one pulse, the echo, is reflected back toward the transducer and the other penetrates into the other material, as shown in the above figure. The brightness of a structure in an ultrasound image depends on the strength of the reflection, or echo. This in turn depends on how much the two materials differ in terms of acoustic impedance. The amplitude ratio of the reflected to the incident pulse is related to the tissue impedance values by

$$\text{Reflection loss (dB)} = 20 \log (Z_2 - Z_1)/(Z_2 + Z_1)$$

At most soft tissue interfaces, only a small fraction of the pulse is reflected. Therefore, the reflection process produces relatively weak echoes. At interfaces between soft tissue and materials such as bone, stones, and gas, strong reflections are produced. The reduction in pulse amplitude during reflection at several different interfaces is given in the following table.

Pulse Amplitude Loss Produced by a Reflection	
Interface	**Amplitude Loss (dB)**
Ideal reflector	0.0
Tissue-air	-0.01
Bone-soft tissue	-3.8
Fat-Muscle	-20.0
Tissue-water	-26.0
Muscle-blood	-30.0

The amplitude of a pulse is attenuated both by absorption and reflection losses. Because of this, an echo returning to the transducer is much smaller than the original pulse produced by the transducer.

Refraction

When an ultrasound pulse passes through an interface at a relatively small angle (between the beam direction and interface surface), the penetrating pulse direction will be shifted by the refraction process.

Pulse Diameter and Beam Width

An important characteristic of an ultrasound pulse is its diameter, which is also the width of the ultrasound beam. The diameter of a pulse changes as it moves along the beam path. At this point we will observe the change in pulse diameter as it moves along the beam and show how it can be controlled.

The diameter of the pulse is determined by the characteristics of the transducer. At the transducer surface, the diameter of the pulse is the same as the diameter of the vibrating crystal. As the pulse moves through the body, the diameter generally changes. This is determined by the focusing characteristics of the transducer.

Transducer Focusing

Transducers can be designed to produce either a focused or non-focused beam, as shown in the following figure. A focused beam is desirable for most imaging applications because it produces pulses with a small diameter which in turn gives better visibility of detail in the image. The best detail will be obtained for structures within the focal zone. The distance between the transducer and the focal zone is the focal depth.

Unfocused Transducers

An unfocused transducer produces a beam with two distinct regions, as shown in the previous figure. One is the so-called near field or Fresnel zone and the other is the far field or Fraunhofer zone. In the near field, the ultrasound pulse maintains a relatively constant diameter that can be used for imaging. In the near field, the beam has a constant diameter that is determined by the diameter of the transducer. The length of the near field is related to the diameter, **D**, of the transducer and the wavelength, **l**, of the ultrasound by

$$\text{Near field length} = D^2/4l$$

Recall that the wavelength is inversely related to frequency. Therefore, for a given transducer size, the length of the near field is proportional to frequency. Another characteristic of the near field is that the intensity along the beam axis is not constant; it oscillates between maximum and zero several times between the transducer surface and the boundary between the near and far field. This is because of the interference patterns created by the sound waves from the transducer surface. An intensity of zero at a point along the axis simply means that the sound vibrations are concentrated around the periphery of the beam. A picture of the ultrasound pulse in that region would look more like concentric rings or "donuts" than the disk that has been shown in various illustrations.

The major characteristic of the far field is that the beam diverges. This causes theultrasound pulses to be larger in diameter but to have less intensity along the central axis. The approximate angle of divergence is related to the diameter of the transducer, **D**, and the wavelength, **l,** by

$$\text{Divergence angle (degrees)} = 70l/D$$

Because of the inverse relationship between wavelength and frequency, divergence is decreased by increasing frequency. The major advantage of using the higher ultrasound frequencies (shorter wavelengths) is that the beams are less divergent and generally produce less blur and better detail.

The previous figure is a representation of the ideal ultrasound beam. However, some transducers produce beams with side lobes. These secondary beams fan out around the primary beam. The principal concern is that under some conditions echoes will be produced by the side lobes and produce artifacts in the image.

Fixed Focus

A transducer can be designed to produce a focused ultrasound beam by using a concaved piezoelectric element or an acoustic lens in front of the element. Transducers are designed with different degrees of focusing. Relatively weak focusing produces a longer focal zone and greater focal depth. A strongly focused transducer will have a shorter focal zone and a shorter focal depth.

Fixed focus transducers have the obvious disadvantages of not being able to produce the same image detail at all depths within the body.

Adjustable Small focus

The focusing of some transducers can be adjusted to a specific depth for each transmitted pulse. This concept is illustrated in the following figure. The transducer is made up of an array of several piëzoelectric elements rather than a single element as in the fixed focus transducer. There are two basic array configurations: linear and annular. In the linear array the elements are arranged in either a straight or curved line. The annular array transducer consists of concentric transducer elements as shown. Although these two designs have different clinical applications, the focusing principles are similar.

Principle of Electronic Focusing with an Array Transducer

Focusing is achieved by not applying the electrical pulses to all of the transducer elements simultaneously. The pulse to each element is passed through an electronic delay. Now let's observe the sequence in which the transducer elements are pulsed in the figure above. The outermost element (annular) or elements (linear) will be pulsed first. This produces ultrasound that begins to move away from the transducer. The other elements are then pulsed in sequence, working toward the center of the array. The centermost element will receive the last pulse. The pulses from the individual elements combine in a constructive manner to create a curved composite pulse, which will converge on a focal point at some specific distance (depth) from the transducer.

The focal depth is determined by the time delay between the electrical pulses. This can be changed electronically to focus pulses to give good image detail at various depths within the body rather than just one depth as with the fixed focus transducer. One approach is to create an image by using a sequence of pulses, each one focused to a different depth or zone within the body.

One distinction between the two transducer designs illustrated here is that the annular array focuses the pulse in two dimensions whereas the linear array can only focus in the one dimension; that is, in the plane of the transducer.

Dynamic Receive Focus

The focusing of an array transducer can also be changed electronically when it is in the echo receiving mode. This is achieved by processing the electrical pulses from the individual transducer elements through different time delays before they are combined to form a composite electrical pulse. The effect of this is to give the transducer a high sensitivity for echoes coming from a specific depth along the central axis of the beam. This produces a focusing effect for the returning echoes.

An important factor is that the receiving focal depth can be changed rapidly. Since echoes at different depths do not arrive at the transducer at the same time, the focusing can be swept down through the depth range to pick up the echoes as they occur. This is the major distinction between dynamic or sweep focusing during the receive mode and adjustable transmit focus. Any one transmitted pulse can only be focused to one specific depth. However, during the receive mode, the focus can be swept through a range of depths to pick up the multiple echoes produced by one transmit pulse.

Conclusion

The ultrasound image is produced by interactions of ultrasound pulses with the anatomical structures within the human body. The basic B mode image is a display of echoes or reflections from the structures and objects. The absorption of the ultrasound as it passes into and back out of the body is generally undesirable because it limits the depth of imaging, adversely affects the amplitude of echoes that form the image, and can be the source of artifacts. The size of the ultrasound pulse at different depths within the imaged area determines the amount of blurring and image detail. An understanding of the physical characteristics of ultrasound and how it interacts with the body enhances the ability to analyze images and make accurate diagnostic decisions.

4

Ultrasonics : Fundamental and Applications

S. K. Shrivastava, Kailash,
Rahul Gupta and Sachin Shrivastava

Introduction

The ultrasonic power supply (generator) converts DC voltage to high frequency 25 kHz (25,000 cylces per second) electrical energy. This electrical energy is transmitted to the transducer within the hand piece, where it is changed to mechanical vibrations. The vibrations from the transducer are intensified by the probe (horn), creating pressure waves in the liquid. This action forms millions of microscopic bubbles (cavities) which expand during the negative pressure excursion, and implode violently during the positive excursion. It is this phenomenon, referred to as cavitation, which produces the powerful shearing action at the probe tip, and causes the molecules in the liquid to become intensely agitated.

Frequency and Amplitude

The radiating-wave frequencies most commonly used in ultrasonic cleaning, 18-120 kHz, lie just above the audible frequency range. In any sonic system,the harmonics of the fundamental frequency, together with vibrations originating at the container walls and liquid surface, produce audible sound. Thus, an operating system that is fundamentally ultrasonic will nonetheless by audible, and low frequency (20-kHz) systems will generally be noisier than higher-frequency (40-kHz) systems.

Moreover, ultrasonic intensity is an integral function of the frequency and amplitude of a radiating wave; therefore, a 20-kHz radiating wave will be approximately twice the intensity of a 40-kHz wave for any given average power output, and consequently the cavitation intensity resulting from a 20-kHz wave will be proportionately greater than that resulting from a 40-kHz wave. The cavitation phenomenon will, of course, occur less frequently at 20 kHz, but this is not thought to have a significant bearing on effectiveness. However, the longer wavelengths of low-frequency ultrasonic systems result in substantially different standing-wave patters throughout the liquid medium. The standing or stationary waves produced by ultrasonics in liquid media result from the simultaneous transmission of the surface-reflected wave motion and the wave motion originating at the transducer radiating surface. The fixed points of minimum amplitude are called nodes, and the points of maximum amplitude are called loops. The distance between the nodes and loops of the 20-kHz standing wave (2 in.) will be approximately twice that of the 40-kHz wave. Because cavitation takes place primarily at the loops, the distance between cavitation sites will thus be larger with 20-kHz than with 40-kHz radiation, and the 20-kHz waves will also have larger dead zones (i.e., zones with little or no cavitation activity). It is for this reason that work resulting from 20-kHz radiation is likely to be less homogeneous and less consistent, even though this frequency produces more intense cavitation. Much of the inhomogeneity in ultrasonic fields can, however, be reduced or wholly eliminated through the use of sweep frequencies, or radiating waves with a multitude of different frequencies. By this means, several overlapping standing waves can be generated at the same time, thereby eliminating much of the dead zone. The amplitude of the radiating wave is directly proportional to the electrical energy that is applied to the transducer. In order for cavitation to be produced in a liquid medium, the amplitude of the radiating wave must have a certain minimum value, which is usually rated in terms of electrical input power to the transducer. No cavitation can occur below this threshold value, and the use of electrical power over and above the minimum level results not in more intense cavitation activity but rather in an increase in the overall quantity of cavitation bubbles. The minimum power requirement for the production of cavitation varies greatly with the colligative properties and temperature of the liquid and with the nature and concentration of dissolved substances.

Cavitation

If a sound wave is impressed upon a liquid and the intensity is increased, a point will be reached where cavitation occurs. Cavitation is the formation of a gas bubble in the liquid during the rarefaction cycle. When the compression cycle occurs the gas bubble collapses. During the collapse tremendous pressures are produced. The pressure may be of the order of several thousand atmospheres. Thousands of these small bubbles are formed in a small volume of the liquid. It is quite generally agreed that it is cavitation that produces most of the biological, detergent, mechanical, and chemical effects in the application of high intensity sound to various mediums. The intensity with which cavitation takes place in a liquid medium varies greatly with the colligative properties of that medium, which include vapor pressure, surface tension, viscosity, and density, as well as any other property that is related to the number of atoms, ions, or molecules in the medium. In ultrasonic cleaning applications, the surface tension and the vapor pressure characteristics of the cleaning fluid play the most significant roles in determining cavitation intensity and, hence, cleaning effectiveness. The energy required to form a cavitation bubble in a liquid is proportional to both surface tension and vapor pressure. Thus, the higher the surface tension of a liquid, the greater will be the energy that is required to produce a cavitation bubble and, consequently, the greater will be the shock-wave energy that is produced when the bubble collapses. In pure water, for example, whose surface tension is about 72 dyne/cm, cavitation is produced only with great difficulty at ambient temperatures. It is, however, produced with facility when a surface-active agent is added to the liquid, thus reducing the surface tension to about 30 dyne/cm. In the same manner, when the vapor pressure of a liquid is low, as is the case with cold water, cavitation is difficult to produce but becomes less and less so as temperature is increased. Every liquid, in fact, has a characteristic/temperature relationship in which cavitation exhibits maximum activity within a fairly narrow temperature range.

Dispersion Due to Ultrasonics

Dispersion in chemistry means the breaking down of a liquid or solid particle into smaller sizes or finer texture and distributing them in another medium. In Chemistry the term system is applied to the whole mixture. Each of the substances comprising the system is called a component. A mixture of two substances is termed a two-component system. The form in which the component exists is called a phase, as, for example, gas, liquid, or solid. A colloidal solution is a two-component system in which a finely divided substance is uniformly distributed through the other. These systems may be classified according to the fineness of dispersion, as, for example, mechanical suspensions, colloidal solutions, and molecular solutions. A list of two-component systems is given below: Solid-Solid, Solid-Liquid, Solid-Gas, Liquid-Solid, Liquid-Liquid, Liquid -Gas, Gas- Solid and Gas-Liquid. All of the above may be obtained in colloidal dimensions. However, liquid + solid and liquid + liquid have received the most attention. Intense sound fields may be used to bring about mixtures in the above systems.

Emulsification Due to Ultrasonics

An emulsion is a suspension of fine particles or globules of a liquid in a liquid. Emulsions are generally produced by violent agitation. This suggests that ultrasonics may be used to produce emulsions. If two immiscible liquids, such as water and gasoline, are placed in a container and subjected to intense sound vibrations it has been found than an emulsion will be formed. The action of ultrasonics in producing emulsification can also be applied to the production of alloys of iron and lead, aluminum and lead, aluminum and cadmium, etc., which are not miscible in the liquid state. It is possible to keep the metals mixed by the application of ultrasonics up to the point of solidification. New bearing materials have been made in this way. Ultrasonics have also been applied to photographic emulsions with an improvement in homogeneity and stability. Ultrasonics have been applied to molten zinc, tin, and aluminum. It was found that the solidification occurred more quickly. In addition, the structure in the solidified state was found to be finer grained. The homogenization of milk, that is, the reduction in size of the fact particles so that cream does not form while the milk stands, can be carried out by means of the application of ultrasonics.

Coagulation Due to Ultrasonics

In spite of the fact that ultrasonics have strong dispersive effects on liquid emulsions their effect on gas and solids and gas and liquids is the opposite - namely, coagulation. The solid and liquid particles in mist, dust, and smoke agglomerate when these mixtures are subjected to intense sound waves. the particles in a small smoke attack have been coagulated and precipitated. The action depends in some degree on the wavelength and intensity. Degassing of molten metals by the application of ultrasonics is another example of coagulation. Small bubbles form at first which join to form larger ones. The larger ones rise to the surface and are expelled. This use of ultrasonics should lead to an improvement in castings where the presence of bubbles is very objectionable.

Chemical Effects of Ultrasonics

A large number of experiments have been conducted on the effect of intense sound waves upon chemical reactions. Certain types of chemical reactions have been speeded by the application of intense waves. However, in some cases it is difficult to isolate the thermal effects due to the sound and the effects due to the sound alone. Another chemical effect is the breaking down of molecules. For example, a chain molecule of starch has been broken into six fragments. The application of intense sound waves to speed up the aging of whiskey has been suggested. the explanation is that in the aging process there is a gradual change in the structure of complex molecules which could be accomplished in a relatively short time with the application of sound.

Biological Effects of Ultrasonics

Ultrasonics have a very destructive effect upon small living organisms. Small fish have been killed by high-power echo ranging and sounding devices. Ultrasonics have been used in the extraction of antigens secreted in the cells of pathogenic bacteria. These antigens are used in serums for immunization against typhoid and other diseases. The bacterial cell walls are broken down by the application of ultrasonic waves and the antigens are set free. The cell walls of the bacteria are separated from the antigens by centrifuging. It appears that bacteria can be destroyed by ultrasonics. The bacteria in milk have been reduced by the application of ultrasonics. This indicates that milk can be sterilized by ultrasonics. Another application in medicine is the use of sound to produce stimulation within the body. Therapeutic effects of a different nature but similar to those produced by heat and radio-frequency diathermy may be obtained. As in the case of chemical effects the biological effects are somewhat obscure but very interesting.

Medical Applications of Ultrasonics

The applications of ultrasonics in the medical field have involved analysis and treatment. The developments in the medical field appear to be very promising. The effect of ultrasonics on tissues has been investigated. The heating and mechanical effects have been isolated. The conclusion is that there is an effect outside of the heating effect. The effects of the changes produced by high intensity sound upon the central nervous system has been investigated. The results show that nerve cells are particularly sensitive to ultrasonics, while blood vessels and nerve fibers are much more resistant. A study of the therapeutic effect of ultrasonics shows that the heat which is produced plays the major role. However, ultrasonics also produces a mechanical effect. Ultrasonics has been used to produce deep-seated heating in the treatment of arthritis. The cerebral ventricular geometry has been portrayed by means of ultrasonic techniques. The head is immersed in water. An underwater projector sends an ultrasonic wave through the head. A hydrophone picks up the transmitted sound. A frequency of 2.5 megacycles was used. A scanning system together with a facsimile-type recorder presents the ultrasonogram in the form of a picture showing the cerebral ventricular geometry. This method provides a means for the detection of brain tumors similar to that of the X-ray. Recent work on tumor detection employs ultrasonic waves and echo-ranging techniques with cathode-ray presentation. the pulses are sent into the body and the echos return in different intensities depending upon the difference in acoustical impedance of the malignant and nonmalignant tissues and in different times depending upon the depths of the reflecting boundaries. A small version of the ultrasonic drill has been developed for use by dentists in drilling teeth.

The advantages of the ultrasonic drill is reduction in pain and improved definition of the drilled area. Destructive ultrasonic probes have been developed for medical use. These probes typically operate between 20 and 60 kilohertz, usually of the piezoelectric type and have high strokes at their distal ends. One such probe is tubular in shape with a small frontal surface area. This surface area produces little or no cavitation when placed in water, however, due to high mechanical vibration this probe can cut through tissue and aspirate the emulsified particles through the center without damage to connective tissue. Another tubular probe widely used for medical applications is the phacoemulsifier. The phacoemulsifier is used in ophthalmology for removing cataract lenses from ones eye. The same principal of mechanical action with little or no cavitation is present at the probes tip that actually cores the lens and aspirates it through the center. Still another medical use of ultrasonics is cutting through tissue with a knife edge. Advantages of this technique for the patient is reduced bleeding from coagulation of blood vessels and the ease at which the knife blade cuts from reduced friction and increased sharpness.

Thermal Effects of Ultrasonics

There is considerable temperature rise in the ultrasonic field in a liquid. A rise of several degrees per minute can be obtained.

The generation in heat is due to dissipation of the sound by absorption in the liquid. The generation of head by the action of ultrasonics obscures the effects which can be attributed to sound alone because many chemical and biological phenomena observed when ultrasonics are applied are also obtained by the application of heat. The practical value of heating by ultrasonics remains to be seen.

Ultrasonics As a Detergent

Ultrasonics may be used to clean and wash various substances. Tests have been made of a ultrasonic washing machine in which clothes mixed with the conventional water and soap solutions are subjected to high intensity sound waves. It has been found that clothes can be cleaned as effectively in this way as by conventional means.

Ultrasonics Cleaning and Degreasing

The use of ultrasonics for cleaning and degreasing surfaces has bound widespread use in industry. Cavitation reduces the surface tension of the clinging dirt and thereby produces a cleaning action on all surfaces and recesses. Cavitation emulsifies greases and oils and thereby assists in the removal of such coatings.

Ultrasonic Drilling

The drilling of glass, ceramics, and metals is now being done by means of ultrasonics. The tip sets up cavitation in a surrounding liquid-borne abrasive slurry. The forces produced by the cavitation bubbles propel the abrasive slurry against the material being drilled. The result is that glass, ceramics, and metals are penetrated in the matter of a few seconds. The point may be any shape as contrasted circular drills. The ultrasonic drill employs a magnetostriction transducer. The system operates as a half-wave resonator. The dimensions of the resonator are usually such that resonance occurs at 20 kilocycles. The amplitude at the drilling tip is increased by the use of a mechanical transformer in the form of a tapered rod.

Ultrasonic Soldering

Aluminum is very difficult to solder because of the oxide which is formed on the surface. When cavitation is induced in the molten solder applied to the surface, the resultant forces break down the metal oxides formed on the surface of the parts being soldered. In this way, the solder is exposed to the pure base metal in a non- oxidizing atmosphere. The ultrasonic energy may be applied to a pot of molten solder or to the tip of a soldering iron. Aluminum may be soldered by dipping the parts in the molten solder in the pot or by the application of the soldering iron. The soldering of aluminum may be carried out without the use of flux.

Testing of Materials by Means of Ultrasonics

A number of systems have been devised for testing materials, particularly metals, for flaws such as hollows, cracks, or other defects of homogeneity. One of the methods employs the distortion of sand patterns on a steel plate when it is caused to vibrate under the influence of sound. This system can only be applied to plates in which the sand pattern is known for a perfect plate. Another system which is particularly useful in that it can be used to detect flaws in a piece of metal of almost any shape is analogous to the echo ranging or depth sounding devices. A quartz crystal projector hydrophone is placed in intimate contact with the metal object to be tested by using a film of oil between crystal and metal. A short pulse of very high-frequency sound (5 megacycles) is sent out by the crystal used as a projector. The reflected pulse is picked up by the crystal used as a hydrophone. The output of the hydrophone is amplified and applied to the screen of a cathode-ray tube. Since all these operations take place in fractions of milliseconds, the electronic switching, etc., is quite intricate and complex. the cathode ray depicts the outgoing pulse and all reflected pulses. From the dimensions and geometry of the piece under test and the velocity of sound in the material and pattern on the oscilloscope it is possible to determine the presence or absence of flaws. This is a very useful and powerful tool. It possesses advantages over X-ray testing in that the particular piece to be tested need not be moved to the apparatus to be tested since the test equipment is quite small and portable. Furthermore, other intervening or adjacent components need not be removed to carry out the tests. A system for detecting flaws in tires by the use of ultrasonics has been developed. The tire is immersed in water and the transmission of an ultrasonic wave through the tire is obtained by a projector and hydrophone combination. since the characteristic acoustical impedance of rubber and water is practically the same, there will be very little attenuation or other anomalies in the transmission of the ultrasonic wave except in the case of a flaw or defect in the rubber.

Ultrasonic Delay Lines and Filters

Delay lines for the storage of pulses one microsecond in length and for periods up to 2000 microseconds have been developed. These delay lines are used for the storage of radar pulses from one pulse to the next. Both mercury and solid lines have been used. Quartz crystal transducers are used for the transmitter and the receiver. Ultrasonic band-pass filters for use in the intermediate frequency amplifiers in radio receivers consist of mass and compliance elements. Magnetostriction transducers are used for the transmitter and the receiver. The outstanding characteristic is the very high attenuation over a very narrow frequency range at the upper and lower cutoff frequencies. For example, a band-pass filter with a pass band of 6 kilocycles at 100 kilocycles shows 45 decibels attenuation in 1 kilocycle at the cutoff frequency.

How Ultrasonic Nozzles Work

Every ultrasonic nozzle operates at a specific resonant frequency, which is determined primarily by the length of the nozzle. In order to produce standing, sinusoidal longitudinal waves, a necessity for the sustained vibration that produces atomization, the nozzle must be an integral number of half-wavelengths long. This requirement arises because both free ends of a nozzle must be anti-nodes; that is, points of maximum vibrational amplitude. Open-ended organ pipes and chimes are other examples of this type of wave motion. The significantly greater amplitude of the standing wave at the atomizing surface end of the nozzle is the result of the amplification of motion provided by the step diameter transition between the large central section of the nozzle and the slender stem that terminates in the atomizing surface.

Principles of Ultrasonic Cleaning

In general, ultrasonic cleaning consists of immersing a part in a suitable liquid medium, agitating or sonicating that medium with high-frequency (18 to 120 kHz) sound for a brief interval of time (usually a few minutes), rinsing with clean solvent or water, and drying. The mechanism underlying this process is one in which microscopic bubbles in the liquid medium implode or collapse under the pressure of agitation to produce shock waves, which impinge on the surface of the part and, through a scrubbing action, displace or loosen particulate matter from that surface. The process by which these bubbles collapse or implode is known as capitation. High intensity ultrasonic fields are known to exert powerful forces that are capable of eroding even the hardest surfaces. Quartz, silicon, and alumina, for example, can be etched by prolonged exposure to ultrasonic cavitation, and "cavitation burn" has been encountered following repeated cleaning of glass surfaces. The severity of this erosive effect has, in fact, been known to preclude the use of ultrasonics in the cleaning of some sensitive, delicate components. Ultrasonic cleaning has, however, been used to great advantage for extremely tenacious deposits, such as corrosion deposits on metals. In any case, cavitation forces can be controlled; thus, given proper selection of critical parameters, ultrasonics can be used successfully in virtually any cleaning application that requires removal of small particulates. Although the ultrasonic cleaning process has been used for over half a century, no reliable means of quantifying its cavitation activity has ever been developed. Indirect methods of measurement, such as erosion tests on metal surfaces, soil removal from weighted samples, acceleration of chemical reactions, thermodynamic studies, and white noise measurement, have been employed to a limited extent, but none of these methods has proved to be effective.

Cleaning Fluid

It is essential that cleaning fluids be selected on the basis of:

- The chemical and physical nature of the contaminants to be removed; and
- The identity of the substrate material.

Insoluble particulate contaminants can, for example, be divided into two groups:

- Water-wettable or hydrophilic particles, including metal particles, metal oxides, minerals, and inorganic dusts.
- Non-water-wettable or hydrophobic substances, including plastic particles, smoke and carbon particles, graphite dust, and organic chemical dusts.

Substrate surfaces, too, can be divided into hydrophilic and hydrophobic groups. Rarely are hydrophobic contaminants found on hydrophilic substrates or vice versa, but when this is the case, cleaning is best accomplished simply through rinsing with a suitable solvent. Hydrophilic particles on hydrophilic substrates, on the other hand, are best removed with aqueous detergent solutions, while hydrophobic particles on hydrophobic substrates are most effectively removed by the use of organic solvents.

Conclusion

From the above discussion, one can, undoubtedly, say that ultrasound finds extensive applications in different aspects of modern life. The latest developments have greatly increased the interest in Ultrasonics and have made the subject to be an active one. The study of ultrasonics, at present, extends an unlimited field of activity for investigators and it opens up immense opportunities for their applications.

References

1. **C.B. Sruby and L.B. Drain**, "*Laser Ultrasonics*", Adam Hilger, UK, 1990.
2. **D.O. Thompson and D.E. Chementi**, "*Review of Progress in Quantitative Non-destructive Evaluation*", 12, 495, 1993.
3. **G. Golan and C. W. Pitt.**, "*Appl. Surface Sci.*", 106, 491, 1996.

4. **R.S. Gilmore,** *J. Appl. Phys.*, 29, 1389, 1996.
5. **J. Kushibiki, T. Okuzawa and Y. Ohashi,** "*J. Appl. Phys.*", 87, 4395, 2000.
6. **J.F. Green Leaf,** "*Science*", 280, 82-85, 1998.
7. **J.F. Green Leaf,** "*Pro. Nail. Acad. Sci.*", USA. 96, 6603, 1999.
8. **K. Pradbhakar and S.P. M. Rao,** "*IAPT Bulletin*", 20, 113, 2003.
9. **P.A. Fokker, J.I. Dijkhuis and H.W. Dewijin,** "*Phys. Rev.*", B55, 2925, 1997.
10. **J.Y. Prieur et al,** "*Physica B.*", 219-220, 235, 1996.
11. **S.S. Makker et al,** "*J. Phys. Condens. Matter*", 10, 5905, 1998.
12. **V. Volkov, S.T. Zavtrak and I.S. Kutten,** "*Phys. Rev.*", 56, 1097, 1997.
13. **P.N.T. Wells,** "*Biomedical Ultrasonics*", Academic Press, London, 1977.
14. **W.L. Nyborg,** "*Ultrasound: Medical Appl.*", Biological elfecs and Hazard Potential, Plenum, N. Y., 1987.
15. **S.K. Shrivastava and Kailash,** "1. *Biosc.*", 30, 269, 2005.
16. **S.K. Shrivastava and Kailash,** "*Bull. Mater. Sci.*", 27, 383, 2004.
17. **M.P. Kapoor,** "*J. Pure Appl. Ultra.*", 19, 104, 1997.
18. **R.K. Saba. S.K. Sen., S.K. Sharma and B.D. Roy,** "*Abstracts*", XII-NSU, Amritsar, 43, 2003.
19. **S. Sal1a, S. Karmakar, R.K. Saha, M. Roy, S. Sarkar and S.K. Sen,** "*Abstracts*", XII-NSU, Amritsar, 44, 2003.
20. **S. Chiled, C.L. Hartman and L.A. Scheryand,** "*Ultr. Med. Bio.*", 16, 817, 1990.
21. **B.B. Goldbergand P.N.T. Walls (Eds),** "*Ultrasonics in Clinical Diagnosis*", Churchill Livingstone, Edinburgh, 1983.
22. **L.A. Frizzel,** "*Encyclopedia of Applied Physics*", VCH Pub., New York, 22, 475, 1998.
23. **P.N.T. Wells,** "*Rep. Prog. Phys.*", 62, 671, 1999.
24. **R.S. Khandpur,** "*Abstracts*", XIl-NSU, Amritsar, 41, 2003.
25. **A.J. Colemon and J.E. Saunders,** "*Ultra. Med. Bio.*", 15, 213, 1980.

5

Medical Ultrasonography

S. K. Shrivastava, Kailash,
Amit Kumar and Ashish Nautiyal

Diagnostic sonography (ultrasonography) is an ultrasound-based diagnostic imaging technique used for visualizing subcutaneous body structures including tendons, muscles, joints, vessels and internal organs for possible pathology or lesions. Obstetric sonography is commonly used during pregnancy and is widely recognized by the public. In physics, the term "ultrasound" applies to all sound waves with a frequency above the audible range of human hearing, about 20,000 Hz. The frequencies used in diagnostic ultrasound are typically between 2 and 18 MHz.

Diagnostic Applications

Orthogonal planes of a 3 dimensional sonographic volume with transverse and coronal measurements for estimating fetal cranial volume. Urinary bladder (black butterfly-like shape) and hyperplastic prostate (BPH) visualized by Medical ultrasonography technique. Typical diagnostic sonographic scanners operate in the frequency range of 2 to 18 megahertz, though frequencies up to 50-100 megahertz has been used experimentally in a technique known as biomicroscopy in special regions, such as the anterior chamber of eye. The choice of frequency is a trade-off between spatial resolution of the image and imaging depth: lower frequencies produce less resolution but image deeper into the body. Higher frequency sound waves have a smaller wavelength and thus are capable of reflecting or scattering from smaller structures.

Higher frequency sound waves also have a larger attenuation coefficient and thus are more readily absorbed in tissue, limiting the depth of penetration of the sound wave into the body.

Sonography (ultrasonography) is widely used in medicine. It is possible to perform both diagnosis and therapeutic procedures, using ultrasound to guide interventional procedures (for instance biopsies or drainage of fluid collections). Sonographers are medical professionals who perform scans for diagnostic purposes. Sonographers typically use a hand-held probe (called a transducer) that is placed directly on and moved over the patient.

Sonography is effective for imaging soft tissues of the body. Superficial structures such as muscles, tendons, testes, breast and the neonatal brain are imaged at a higher frequency (7-18 MHz), which provides better axial and lateral resolution. Deeper structures such as liver and kidney are imaged at a lower frequency 1-6 MHz with lower axial and lateral resolution but greater penetration. Medical sonography is used in the study of many different systems:

Other types of uses include:

- Intervenional; biopsy, emptying fluids, intrauterine transfusion (Hemolytic disease of the newborn)
- Contrast-enhanced ultrasound

A general-purpose sonographic machine may be used for most imaging purposes. Usually specialty applications may be served only by use of a specialty transducer. Most ultrasound procedures are done using a transducer on the surface of the body, but improved diagnostic confidence is often possible if a transducer can be placed inside the body. For this purpose, specialty transducers, including endovaginal, endorectal, and transesophageal transducers are commonly employed. At the extreme of this, very small transducers can be mounted on small diameter catheters and placed into blood vessels to image the walls and disease of those vessels.

Therapeutic Applications

Therapeutic applications use ultrasound to bring heat or agitation into the body. Therefore much higher energies are used

System	***Description***
Anesthesiology	Ultrasound is commonly used by anesthesiologists (Anaesthetists) to guide injecting needles when placing local anaesthetic solutions near nerves
Cardiology	Echocardiography is an essential tool in cardiology, to diagnose e.g. dilatation of parts of the heart and function of heart ventricles and valves
Emergency Medicine	Point of care ultrasound has many applications in the Emergency Department, including the Focused Assessment with Sonography for Trauma (FAST) exam for assessing significant hemoperitoneum or pericardial tamponade after trauma. Ultrasound is routinely used in the Emergency Department to expedite the care of patients with right upper quadrant abdominal pain who may have gallstones or cholecystitis.
Gastroenterology	In abdominal sonography, the solid organs of the abdomen such as the pancreas, aorta, inferior vena cava, liver, gall bladder, bile ducts, kidneys, and spleen are imaged. Sound waves are blocked by gas in the bowel and attenuated in different degree by fat, therefore there are limited diagnostic capabilities in this area. The appendix can sometimes be seen when inflamed (as in e.g.: appendicitis).
Gynecology	
Neonatology	for basic assessment of intracerebral structural abnormalities, bleeds, ventriculomegaly or hydrocephalus and anoxic insults (Periventricular leukomalacia). The ultrasound can be

	performed through the soft spots in the skull of a newborn infant (Fontanelle) until these completely close at about 1 year of age and form a virtually impenetrable acoustic barrier for the ultrasound. The most common site for cranial ultrasound is the anterior fontanelle. The smaller the fontanelle, the poorer the quality of the picture.
Neurology	for assessing blood flow and stenoses in the carotid arteries (Carotid ultrasonography) and the big intracerebral arteries
Obstetrics	Obstetrical sonography is commonly used during pregnancy to check on the development of the fetus.
Ophthalmology	
Urology	to determine, for example, the amount of fluid retained in a patient's bladder. In a pelvic sonogram, organs of the pelvic region are imaged. This includes the uterus and ovaries or urinary bladder. Males are sometimes given a pelvic sonogram to check on the health of their bladder, the prostate, or their testicles (for example to distinguish epididymitis from testicular torsion). In young males, it is used to distinguish more benign testicular masses (varicocele or hydrocele) from testicular cancer, which is still very highly curable but which must be treated to preserve health and fertility. There are two methods of performing a pelvic sonography - externally or internally. The internal pelvic sonogram is performed either transvaginally (in a woman) or transrectally (in a man). Sonographic imaging of the pelvic floor can produce important diagnostic information regarding the precise relationship of abnormal structures with

	other pelvic organs and it represents a useful hint to treat patients with symptoms related to pelvic prolapse, double incontinence and obstructed defecation. It is used to diagnose and, at higher frequencies, to treat (break up) kidney stones or kidney crystals (nephrolithiasis).
Musculoskeletal	tendons, muscles, nerves, ligaments, soft tissue masses, and bone surfaces
Cardiovascular system	To assess patency and possible obstruction of arteries Arterial sonography, diagnose DVT (Thrombosonography) and determine extent and severity of venous insufficiency (venosonography)

than in diagnostic ultrasound. In many cases the range of frequencies used are also very different.

- Ultrasound is sometimes used to clean teeth in dental hygiene.
- Ultrasound sources may be used to generate regional heating and mechanical changes in biological tissue, e.g. in occupational therapy, physical therapy and cancer treatment. However the use of ultrasound in the treatment of musculoskeletal conditions has fallen out of favor.
- Focused ultrasound may be used to generate highly localized heating to treat cysts and tumors (benign or malignant), This is known as Focused Ultrasound Surgery (FUS) or High Intensity Focused Ultrasound (HIFU). These procedures generally use lower frequencies than medical diagnostic ultrasound (from 250 kHz to 2000 kHz), but significantly higher energies. HIFU treatment is often guided by MRI.
- Focused ultrasound may be used to break up kidney stones by lithotripsy.
- Ultrasound may be used for cataract treatment by phacoemulsification.

- Additional physiological effects of low-intensity ultrasound have recently been discovered, e.g. its ability to stimulate bone-growth and its potential to disrupt the blood-brain barrier for drug delivery.
- Procoagulant at 5-12 MHz,

Sound to Image

The creation of an image from sound is done in three steps - producing a sound wave, receiving echoes, and interpreting those echoes.

Producing a Sound Wave

A sound wave is typically produced by a piezoelectric transducer encased in a housing which can take a number of forms. Strong, short electrical pulses from the ultrasound machine make the transducer ring at the desired frequency. The frequencies can be anywhere between 2 and 18 MHz. The sound is focused either by the shape of the transducer, a lens in front of the transducer, or a complex set of control pulses from the ultrasound scanner machine (Beamforming). This focusing produces an arc-shaped sound wave from the face of the transducer. The wave travels into the body and comes into focus at a desired depth.

Older technology transducers focus their beam with physical lenses. Newer technology transducers use phased array techniques to enable the sonographic machine to change the direction and depth of focus. Almost all piezoelectric transducers are made of ceramic.

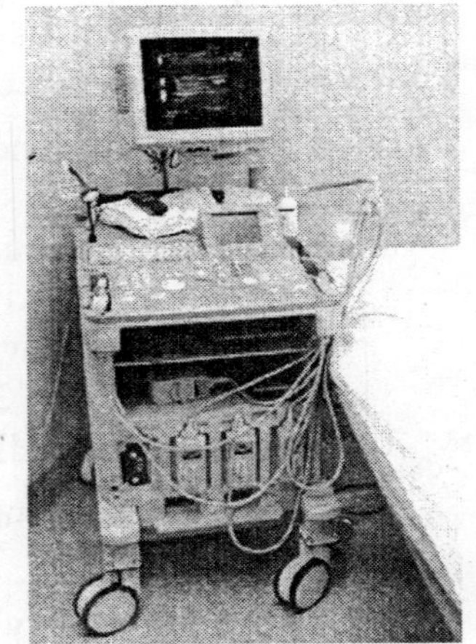

Fig. 1.

Materials on the face of the transducer enable the sound to be transmitted efficiently into the body (usually seeming to be a rubbery coating, a form of impedance matching). In addition, a water-based gel is placed between the patient's skin and the probe.

The sound wave is partially reflected from the layers between different tissues. Specifically, sound is reflected anywhere there are density changes in the body: e.g. blood cells in blood plasma, small structures in organs, etc. Some of the reflections return to the transducer.

Receiving the Echoes

The return of the sound wave to the transducer results in the same process that it took to send the sound wave, except in reverse. The return sound wave vibrates the transducer, the transducer turns the vibrations into electrical pulses that travel to the ultrasonic scanner where they are processed and transformed into a digital image.

Forming the Image

The sonographic scanner must determine three things from each received echo:

- How long it took the echo to be received from when the sound was transmitted.
- From this the focal length for the phased array is deduced, enabling a sharp image of that echo at that depth (this is not possible while producing a sound wave).
- How strong the echo was. It could be noted that sound wave is not a click, but a pulse with a specific carrier frequency. Moving objects change this frequency on reflection, so that it is only a matter of electronics to have simultaneous Doppler sonography.

Once the ultrasonic scanner determines these three things, it can locate which pixel in the image to light up and to what intensity and at what hue if frequency is processed.

Transforming the received signal into a digital image may be explained by using a blank spreadsheet as an analogy. First picture a long, flat transducer at the top of the sheet. Send pulses down the 'columns' of the spreadsheet (A, B, C, etc.). Listen at each column for any return echoes. When an echo is heard, note how long it took for the echo to return. The longer the wait, the deeper the row (1,2,3, etc.). The strength of the echo determines the brightness setting for that cell (white for a strong echo, black for a weak echo, and varying shades of grey for everything in between.) When all the echoes are recorded on the sheet, we have a greyscale image.

Displaying the Image

Images from the sonographic scanner can be displayed, captured, and broadcast through a computer using a frame grabber to capture and digitize the analog video signal. The captured signal can then be post-processed on the computer itself.

Sound in the Body

Ultrasonography (sonography) uses a probe containing multiple acoustic transducers to send pulses of sound into a material. Whenever a sound wave encounters a material with a different density (acoustical impedance), part of the sound wave is reflected back to the probe and is detected as an echo. The time it takes for the echo to travel back to the probe is measured and used to calculate the depth of the tissue interface causing the echo. The greater the difference between acoustic impedances, the larger the echo is. If the pulse hits gases or solids, the density difference is so great that most of the acoustic energy is reflected and it becomes impossible to see deeper. The frequencies used for medical imaging are generally in the range of 1 to 18 MHz. Higher frequencies have a correspondingly smaller wavelength, and can be used to make sonograms with smaller details. However, the attenuation of the sound wave is increased at higher frequencies, so in order to have better penetration of deeper tissues, a lower frequency (3-5 MHz) is used.

Seeing deep into the body with sonography is very difficult. Some acoustic energy is lost every time an echo is formed, but most of it (approximately $0.5\frac{\text{dB}}{\text{cm depth. MHz}}$ is lost from acoustic absorption. The speed of sound varies as it travels through different materials, and is dependent on the acoustical impedance of the material. However, the sonographic instrument assumes that the acoustic velocity is constant at 1540 m/s. An effect of this assumption is that in a real body with non-uniform tissues, the beam becomes somewhat de-focused and image resolution is reduced.

To generate a 2D-image, the ultrasonic beam is swept. A transducer may be swept mechanically by rotating or swinging. Or a 1D phased array transducer may be use to sweep the beam electronically. The received data is processed and used to construct the image. The image is then a 2D representation of the slice into the body. 3D images can be generated by acquiring a series of adjacent 2D images. Commonly a specialised probe that mechanically scans a conventional 2D-image transducer is used. However, since the mechanical scanning is slow, it is difficult to make 3D images of moving tissues. Recently, 2D phased array transducers that can sweep the beam in 3D have been developed. These can image faster and can even be used to make live 3D images of a beating heart. Doppler ultrasonography is used to study blood flow and muscle motion. The different detected speeds are represented in color for ease of interpretation, for example leaky heart valves: the leak shows up as a flash of unique color. Colors may alternatively be used to represent the amplitudes of the received echoes.

Modes of Sonography

Several different modes of ultrasound are used in medical imaging. These are:

A-Mode

A-mode is the simplest type of ultrasound. A single transducer scans a line through the body with the echoes plotted on screen as a function of depth. Therapeutic ultrasound aimed at a specific tumor or calculus is also A-mode, to allow for pinpoint accurate focus of the destructive wave energy.

B-Mode

In B-mode ultrasound, a linear array of transducers simultaneously scans a plane through the body that can be viewed as a two-dimensional image on screen.

C-Mode

A C-mode image is formed in a plane normal to a B-mode image. A gate that selects data from a specific depth from an A-mode line is used; then the transducer is moved in the 2D plane to sample the entire region at this fixed depth. When the transducer traverses the area in a spiral, an area of 100 cm^2 can be scanned in around 10 seconds.

M-Mode

M stands for motion. Ultrasound pulses are emitted in quick succession - each time, either an A-mode or B-mode image is taken. Over time, this is analogous to recording a video in ultrasound. As the organ boundaries that produce reflections move relative to the probe, this can be used to determine the velocity of specific organ structures.

Doppler Mode

This mode makes use of the Doppler effect in measuring and visualizing blood flow.

Color Doppler

Velocity information is presented as a color coded overlay on top of a B-mode image.

Continuous Doppler

Doppler information is sampled along a line through the body, and all velocities detected at each time point is presented (on a time line).

Pulsed Wave (PW) Doppler

Doppler information is sampled from only a small sample volume (defined in 2D image), and presented on a timeline.

Duplex

A common name for the simultaneous presentation of 2D and (usually) PW doppler information. (Using modern ultrasound machines color doppler is almost always also used, hence the alternative name Triplex).

Pulse Inversion Mode

In this mode two successive pulses with opposite sign are emitted and then subtracted from each other. This implies that any linearly responding constituent will disappear while gases with non-linear compressibility stands out.

Harmonic Mode

In this mode a deep penetrating fundamental frequency is emitted into the body and a harmonic overtone is detected. In this way depth penetration can be gained with improved lateral resolution.

Attributes

As with all imaging modalities, ultrasonography has its list of positive and negative attributes.

Strengths

- It images muscle, soft tissue, and bone surfaces very well and is particularly useful for delineating the interfaces between solid and fluid-filled spaces.
- It renders "live" images, where the operator can dynamically select the most useful section for diagnosing and documenting changes, often enabling rapid diagnoses. Live images also allow for ultrasound-guided biopsies or injections, which can be cumbersome with other imaging modalities.
- It shows the structure of organs.
- It has no known long-term side effects and rarely causes any discomfort to the patient.
- Equipment is widely available and comparatively flexible.

- Small, easily carried scanners are available; examinations can be performed at the bedside.
- Relatively inexpensive compared to other modes of investigation, sucn as computed X-ray tomography, DEXA or magnetic resonance imaging.
- Spatial resolution is better in high frequency ultrasound transducers than it is in most other imaging modalities.
- Through the use of an Ultrasound research interface, an ultrasound device can offer a relatively inexpensive, real-time, and flexible method for capturing data required for special research purposes for tissue characterization and development of new image processing techniques

Weaknesses

- Sonographic devices have trouble penetrating bone. For example, sonography of the adult brain is very limited though improvements are being made in transcranial ultrasonography.
- Sonography performs very poorly when there is a gas between the transducer and the organ of interest, due to the extreme differences in acoustic impedance. For example, overlying gas in the gastrointestinal tract often makes ultrasound scanning of the pancreas difficult, and lung imaging is not possible (apart from demarcating pleural effusions).
- Even in the absence of bone or air, the depth penetration of ultrasound may be limited depending on the frequency of imaging. Consequently, there might be difficulties imaging structures deep in the body, especially in obese patients.
- Body habitus has a large influence on image quality, image quality and accuracy of diagnosis is limited with obese patients, overlying subcutaneous fat attenuates the sound beam and a lower frequency tranducer is required (with lower resolution)
- The method is operator-dependent. A high level of skill and experience is needed to acquire good-quality images and make accurate diagnoses.

- There is no scout image as there is with CT and MRI. Once an image has been acquired there is no exact way to tell which part of the body was imaged.

Risks and Side-Effects

Ultrasonography is generally considered a safe imaging modality, although there is relatively little data available. Diagnostic ultrasound studies of the foetus are generally considered to be safe during pregnancy. This diagnostic procedure should be performed only when there is a valid medical indication, and the lowest possible ultrasonic exposure setting should be used to gain the necessary diagnostic information under the "as low as reasonably achievable" or ALARA principle. World Health Organizations technical report supports that ultrasound is harmless: "Diagnostic ultrasound is recognized as a safe, effective, and highly flexible imaging modality capable of providing clinically relevant information about most parts of the body in a rapid and cost-effective fashion". Although there is no evidence ultrasound could be harmful for the foetus, US Food and Drug Administration views promotion, selling, or leasing of ultrasound equipment for making "keepsake foetal videos" to be an unapproved use of a medical device.

References

1. **Portis, J. Andrew and C.P. Sundaram**, "*Am. Family Physician*", 63 (2001) 1329.
2. **H. Bode, M. Sauer and W. Pringsheim**, "*Arch. Dis. Child*", 63 (1988) 1474.
3. **Lei H., Chen and Liu Z.,** *Int. J. Dermatol.*, 49(3) (2010) 311.
4. **Peterson A.,** *Eur. J. Plast. Surg.* 32 (2009) 283.
5. **Suslick and Flannigan K. S.,** An. Rev. Phys. Chem., 59 (2008) 659.
6. **Lee J. and Dong X.,** *Ultras. Sonochem.*, 17 (2010) 649.
7. **Kamarchik E., Fu B. and Bowman,** *J. M., J. Chem. Phys.*, 132 (2010) 091102.

8. **Pandey D. K., Yadawa P. K. and Yadav R. R.,** *Mat. Lett.*, 61 (2007) 5194.
9. **Yadav R. R., Yadawa P. K.. Gupta A. K. and Pandey D. K.,** *J. Acous. Soc. Ind.*, 33 (2005) 219.
10. **Yadav R. R., Mishra G., Yadawa P. K., Kor S. K. Gupta A. K., Raj B. and Jayakumar T.,** *Ultras.*, 48 (2008) 591.
11. **Sivakumar M., Towata A., Yasui K., Tuziuti T. and Iida Y.,** *Curr. Appl. Phys.*, 6 (2006) 591.
12. **Zweifel R. and Zeugin F.,** *New Phytologist*, 179 (2010) 1070.
13. **Telschow K. L., Deason V. A., Cottle D. L. and Larson J. D.,** *IEEE Ultrasonic Symposium, Puerto Rico*, October 22-25 (2000).
14. **Frankel A. S.and Clark C. W.,** *Marine Mamm. Sc.*, 18 (2009) 664.
15. **Cornuelle B. D.,** *J. Phys.: Conf. Ser.* 118 (2008) 12002.
16. **Mobley J. R.,** *J. Acou. Soc. Am.*, 117 (2009) 1666.
17. **Zaijin L., Liming H.and Ye W.,** *J. Semicondu.*, 31 (2010) 036002.
18. **Guzzo P. L.,** "*A Comparative Study on Ultrasonic Machining of Hard and Brittle Materials,*" ABCM, Vol. XXVI, (2004) 56.
19. **Gong H., Fang F. Z. and Hu X. T.,** *Int. J. Machine Tools Manu.*, 50 (2010) 303.
20. **Shrivastava S. K., Kailah, Anchala and Singh J.,** *New Dimensions of Physics*, Crescent Publishers, Allahabad, Ch. 9 (2011) 65.

6

Application of Sound Frequency As a Weapon

S. K. Shrivastava, Kailash
Rajendra Kumar and Anjita Shrivastava

Introduction

The human ear can only hear a limited part of the sound spectrum. Above that range is ultrasound and below it is infrasound; although largely unheard, vibrations in these ranges can still affect the human body in ways that are quite different from the informational aspect of simply listening. These higher and lower registers of sound frequencies are, today, the stuff of imaginative speculation. While the conspiracy watchers believe they are the basis of secret weapons research for covert operations, mind control and other conspiratorial uses, another, more idealistic, school associates them with meditative states and magical technology. The wilder fortean literature attributes to the builders of ancient monuments everywhere the secret of levitating blocks of stone by their mastery of sound; the Vedic gods to power their vimana flying ships supposedly also used such powers.

The use of disconcerting noise to unsettle the enemy is hardwired into most higher animals, from the warnings and battle roars of confrontational beasts to the trumpets, drums, bugles, bagpipes, devilish war cries, taunts and piercing shrieks used by humans in their conflicts. Similarly, during the Gulf War, in the prelude to the final massacre of the fleeing Iraqi forces on the road to Basra, American soldiers were reported to have blasted grunge and death rock from speakers mounted on their vehicles. Yet in these cases, for all the psychological terror the noise was intended to create, it was a crude application of volume and culturally jarring music rather than the directed application of a sound frequency as a weapon.

History

Possibly the earliest account in Western literature of sound itself being used as a weapon can be found in the Bible. As detailed in Joshua 6:5, Joshua leads an attack on the city of Jericho (c1400 BC) during which he commands his people, outside the walled city, to remain in total silence for seven days. On the seventh day, seven trumpets made from ram's horns give a "long blast", the people shout… and the walls of Jericho come crashing down. (It is significant that silence is used as well as noise and perhaps even ultrasound.)

Sound is a waveform, with low infrasonic frequencies having a long wave length (measured in tens of metres), and with high ultrasonic frequencies having a short wave length (measured in millimetres). The frequencies associated with ultrasound are most familiar from their utilization by the medical profession, chiefly for diagnostic imaging. While the ears are designed to detect a limited range of frequencies – the human auditory range is between 20Hz and 20,000Hz (1Hz = 1 cycle per second) – different frequencies can affect the whole body and, at volume, can be felt in almost any part of the body. Even with industrial ear protectors, sound waves are able to enter the head via the nose and mouth which are, in turn, linked to the ears by the structure of the skull. Sounds that are higher in frequency than 20,000Hz – ultrasound – are inaudible to humans, while sounds lower than 20Hz – infrasound – are inaudible but can, on occasion, be felt resonating within the body itself. Exposure of unprotected ears to infrasound can also cause an increase in pressure within the middle ear, disturbing the sense of balance.

Scientists have developed ways of measuring infrasound associated with these phenomena to aid their research. The military use of infrasound dates back to the First World War, when the detection of such frequencies helped pinpoint the enemy's heavy artillery. The idea that infrasound could actually be used as a weapon tends to be attributed to Axis scientists, but of course much of the weapons research by the Axis powers was also of interest to Allied military scientists (see 'Sounds suspicious' panel). The potential of infrasound to affect the human body has long been apparent; as anybody who has leant against the PA at a rave will tell you, even audible sub-bass frequencies at the correct volume can churn your stomach. The theory behind infrasound weapons tends to focus on the idea that certain frequencies can be used as both a weapon and as a method of crowd control.

According to the Working Paper on Infrasound Weapons produced by Hungary for the United Nations in 1978, the frequency that is thought to be most dangerous to humans is between 7 and 8Hz. This is the resonant frequency of flesh and, theoretically, it can rupture internal organs if loud enough. Seven hertz is also the average frequency of the brain's alpha rhythms; thus this frequency has been described as dangerous but also relaxing. Whether exposure to such infrasound can trigger epileptic seizures, as some fear, remains unclear; experimental data on exposure to such frequencies gives a variety of results. It should be noted, however, that the strobe light effect associated with triggering epileptic seizures flashes at an equivalent rhythm. Frequencies below 50Hz commonly lose their coherence and are perceived to pulse or fluctuate, which is analogous to the strobing beat of a modulated light.

It was NASA scientists in the early 1960s who produced most of the documentation of the effects of infrasound on the human body; they were particularly keen to discover how proximity to the low frequencies produced by rocket engines would affect their astronauts, especially during launching. Their extensive tests confirmed that, at certain volumes, infrasound did indeed have various physiological consequences. According to results published by NASA researcher GH Mohr, frequencies between 0Hz and 100Hz, at up to 150-155dB, produced vibrations of the chest wall, changes in respiratory rhythm, gagging sensations, headaches, coughing, visual distortion, and post-exposure fatigue. Subsequent research has determined that the frequency that causes vibration of the eyeballs – and therefore distortion of vision – is around 19Hz.

Tandy believes that 'ghost hunters' could benefit from investigating the infrasound frequencies at other 'haunted' locales. Not only does the 19Hz frequency create visual disturbances by vibrating the eyeball – hence the shimmering appearance of apparitions – but the frequency could also stimulate a psychological sense of disquiet (hairs on the back of the neck rising and so forth). Even the drop in temperature associated with spectral manifestations could be an effect of infrasound. It does not cause a measurable drop in temperature of the air but "the effect is caused by a reaction in the body.

Effects like these could also, theoretically, be contributing to sick-building syndrome as standing waves of infrasound can be created by architectural anomalies or frequencies set up by electronic devices. In 1978, the artist-industrial musician Monte Cazazza and the group Throbbing Gristle (above) experimented in their East London studio with the creation of both ultrasound and infrasound frequencies. Cazazza remembers during infrasound tests using an industrial tone generator that the air began to shimmer and his clothes visibly "rippled under the waves." The group's ultrasound experiments were equally notorious; using an array of piezo-electric speakers ("because they were cheap" remembers Monte), they used frequencies in excess of 20,000Hz in a 'sonic loop', creating a continual, culminating wave. Their target was some troublesome neighbours; according to the group, the neighbours' dogs began to bark and both people and animals exhibited aggressive irritability. Unsurprisingly, the unwanted neighbours moved shortly after the sonic attacks.

There is good reason to believe, then, that exposure to certain infrasound frequencies could stimulate aggression and exacerbate psychological disturbances. This might explain accounts of 'temporary psychosis' associated with some natural phenomena, such as the Mistral (in the Rhone Valley) and the Sirocco (off the Sahara), the famous winds that are said to create periods of momentary insanity. That certain gusts of wind have infrasound frequencies has been documented.

The link between periods of insanity and exposure to specific infrasound frequencies forms the basis for the 'Feraliminal Lycanthropizer', a device claimed to stimulate atavistic animality, sexual excitement, and a loss of inhibitions in its target. The Feraliminal Lycanthropizer creates two infrasound frequencies – 3Hz and 9Hz – which, combined, generate a lower, third frequency of 0.56Hz. The machine also uses a combination of four subliminal, looped, audio tape recordings – playing both forwards and backwards – outside the normal audible pitch. A broadcast infrasound weapon would, indeed, cause more trouble than it is worth. In open air, the energy required to drive it is enormous and the effects unpredictable, ranging from serious harm to very little depending on the individual targeted. Directing infrasound is difficult because of the long wavelength, so if the weapon is to be activated by a person holding it, it would be hard to protect them from the sound. Direct contact with the weapon might also pose vibration problems for the operator.

DEvelopment

There is, however, evidence to suggest military and law enforcement authorities have considered that ultrasound as a likely technology for so-called 'non-lethal weapons' for use in crowd control and 'coercive interrogation'. 'White noise' is believed to have been a key element in sensory deprivation techniques since the early 1970s and ultrasonic riot control devices are also believed to have been deployed in quelling civil unrest. One such device – the 'squawk box' – blasts two slightly different, intolerably high-pitched ultrasound frequencies (16,000Hz and 16,002Hz) at rioters; the two, when combined in the ear, effectively produce the frequencies 32,002Hz and 2Hz.

Predictably, the media image of the use of infrasound is as a weapon that disables the body and discomforts the mind; however, it has also been discussed in association with enlightened meditative states. The mantras and chants of monks, priests and followers of a variety of religions are commonly believed to have a profoundly calming effect on practitioners just as some musical instruments – like Tibetan thigh bone trumpets – are thought to resonate at the same frequency as the human body, whilst Tibetan singing bowls are believed to trigger specific frequencies in the brain. A significant part of this old 'mystical' technology is the ritual buildings (tombs, chambers, cathedrals and temples) designed to amplify or modulate the resonances created by rhythmic chants, singing or music.

The activity in our brains functions at several specific frequencies so it seems logical that certain frequencies of sound which are harmonics of that neural activity may influence brain-specific activities. Audiotapes are available which are designed to stimulate the relaxing frequencies associated with meditative states via a process of binaural beats. These recordings work by sending different frequencies to each ear which, when combined in the brain, produce a therapeutic 'pink noise'. Thus, an 800Hz tone in one ear, and 810Hz in the other, would create a 10Hz frequency intended to soothe the Alpha waves. Whether these binaural tapes work depends, undoubtedly, on the listener's susceptibility to sound and to the philosophy associated with the tapes.

One might not notice it, but infrasound permeates our daily environment; the machines around us, the buildings, and the weather all generate infrasound frequencies. The effects may be as unsettling as a ghostly vision, as tiring as the pressure created before a storm, or as invigorating as a good night's sleep.

Disabling forms of infrasound may be used in future wars or to quell civil riots and demonstrations. With important consequences like these, it is unsettling to realise that we actually know far too little about the audio frequencies that surround us.

During the Second World War, workers in munitions factories would listen to the radio as they worked, and it was observed that they seemed to keep pace with the rhythm of the broadcast music. The faster the beat, the faster the production line would move. Much has been made of 'muzak' and the way in which it has been used to both sooth and motivate people in factories, office buildings and shopping malls. Muzak's pop-derived tones are intended to create – broadly speaking – a relaxed environment. However, muzak can be used as carrier for subliminal (hidden) messages which, for example, dissuade thieves in shops. Cynics have suggested that muzak could also be exploited to convey messages urging greater consumption in shops and increased work in factories. It is almost impossible to tell how successful these anti-crime and pro-shopping messages are, but their continued existence suggests that at least some of those investing in the shopping-as-leisure industry believe they are thereby turning muzak into a global industry.

Sound Suspicious

Any discussion of sonic weapons has to contend with a huge volume of internet-circulated misinformation. As one scientist put it: "One cannot avoid the impression that much of what is written on acoustic weapons is based on hearsay and misunderstandings.

Leaving aside the wilder claims about the German secret weapons programmes of World War II, it is certainly true that scientists under Hitler's regime were involved in projects covering just about every conceivable area of weaponry. The best known were the 'V' weapons and the rocket and jet-propelled fighters like the Me163 and the Me262 – but Allied intelligence, by the end of the war, had uncovered a vast array of far more bizarre projects, the development of which had been encouraged by Germany's non-centralised and chaotic approach to R&D. In the words of one contemporary American intelligence report: "There were more crackpot notions getting political support than we would have imagined." Some of the most eccentric projects seem to have originated with an Austrian researcher called Dr Zippermeyer, whose response to the ferocious Allied air bombardment of the Reich was to experiment with both wind and sound as potential anti-aircraft weapons.

One such device was the Windkanone or 'Whirlwind Cannon' (above) , which was meant to produce artificial 'whirlwinds' by generating explosions in a combustion chamber and directing them through specially designed nozzles at the target. Experiments with a small cannon supposedly shattered planks at 200-yard (183m) range, and a full-size one was built. Fortunately for British and American aircraft, the effect was impossible to reproduce at high altitudes and the project was scrapped. The huge hulk of the 'Whirlwind Cannon' itself, though, was discovered rusting and abandoned by bemused Allied forces on the Artillery Proving ground at Hillersleben in April 1945.

Experimenting with the destructive properties of sound was a logical course for Zippermeyer, whose labs also worked on the Luftkanone or 'Sound Cannon' which burned methane and air to produce a rapid series of explosions that were beamed by 'sound-mirrors' into the sky; the resulting noise built up into a high-pitched tone which, apparantly, had been shown as lethal to animals at close range and uncomfortable for human beings at 300 yards (274m). Ultimately, though, the 'Sound Cannon' was doomed by the same limitations that had beset the 'Whirlwind Cannon' – the impossibility of getting the destructive effects high enough to actually attack a flying target.

To demonstrate the confusion surrounding the whole subject, other accounts speak of a 'Sound Cannon' designed by a Dr Richard Wallauschek, a 'Vortex Gun' attributed to a 'Dr Zimmermayer' and a 'Wind Cannon' built at Stuttgart that was supposedly employed defensively at a bridge on the Elbe. Most of these accounts are unreferenced, and all seem to be more or less imaginative variants on the Zippermeyer devices.

The name most often mentioned in connection with the deadly potential of infrasound is that of French robotics researcher Dr Gavreau (sometimes given as 'Gavraud'), variously credited with having made some significant discoveries "around 1957", "in 1965" and "in the early '70s". To boil the story down to its essentials, Gavreau and his team experienced inexplicable bouts of nausea in their lab. These were eventually traced to a faulty motor-driven ventilator which, with the aid of a large concrete duct, was producing an infrasonic resonance.

References

1. **Schindel D.W., Hutchins D.A., Zou L. and Sayer M.,** *IEEE trans.* Ultrason. Ferroelect. & Freq. Control., 42 (1995) 42.
2. **Ladabaum I., Jin X., Soh H.T., Atalar A. and Khuri-Yakub B.T.,** *IEEE Trans. Ultrason. Ferroelect. & Freq. Control.*, 45 (1998) 678
3. **Rayleigh L.,** *Proc. London Math. Soc.*, 17 (1885) 4.
4. **Du J., Harding G.L., Ogilvy J.A., Dencher P.R. and Lake M.,** *Sensors & Actuators*, A56 (1996) 211.
5. **Olinar A.A.,** *Acoustic Surface Waves, Berlin and New York*, Springer Verleg, (1978).
6. **Love A.E.,** *Some Problems of Geodynamics*, Cambridge U.P., Cambridge, England (1891).
7. **White R.M.,** *Methods of Experimental Physics*, New York, Academic Press (1981) 19, 10.
8. **Vandermeer P.R., Meijer G.C.M., Vellekoop M.J., Kerkvliet H.M. M. and Van Den Boom T.J.J.,** *Sensors & Actuators*, A71 (1998) 27.
9. **Hornsteiner J., Born E., Fischerauer G. and Riha E.,** *Proceedings of the IEEE International Frequency Control Symposium* (1998) 615.
10. **Dickert F.L., Forth P., Gulst W.E., Fischerauer G. and Knauer U.,** *Sensors & Actuators*, B46 (1998) 120.
11. **Chu B.,** *Laser Light Scattering*, New York, Academic Press (1974).
12. **Damon R.W., Maloney W.T. and Mc Mahon D.H.,** *Physical Acoustics*, New York and London, Academic Press (1970) 7, 273.
13. **Bergmann L.,** *Ultrasonics*, London, Bell and Sons (1938).
14. **Fleury P.A.,** *Physical Acoustics*, New York and London, Academic Press (1970) 6, 2.
15. **Parker T.E.,** *IEEE Ultrason. Symp. Proc.*, New york (1974). 365
16. **Sun D.G. and Chem R.T.,** *Appl. Phys. Lett.*, 72 (1998) 3139.

7

Science and Technology of Acoustic Emission

S. K. Shrivastava, Rahu Gupta,
Anchala and Pushpraj Singh

Introduction

Acoustic emission is an amazing, promising and challenging subject of the modern technology and science. It is a well known from everyday life phenomenon: sound of breaking glass, falling tree, cracking ice are some examples of fracture sound we may hear from different objects subjected to stress. Scientifically defined, acoustic emission is a phenomenon of sound and ultrasound wave generation by materials that undergo deformation and fracture processes. Sources generating AE in different materials are unique. For examples, in metals, primary macroscopic sources are crack jumps, processes related to plastic deformation development and fracturing and de-bonding of inclusions. Quantitative and qualitative characteristics of acoustic emission waves, generated by sources of different nature depend directly on material properties and environmental factors. Leaks, friction, knocks, chemical reactions, changes of size of magnetic domains are other examples of sources generating acoustic emission waves. These sources belong to another, secondary class of acoustic emission that is usually distinguished from the primary class of sources related to deformation and fracture development. Understanding the nature of emitted sound, characteristics of sounds and what they represent, can be used for development of useful technological solutions in non-destructive testing, material studies, control of production, medical examinations, analysis of chemical reactions and many other fascinating applications. Presentation of fundamentals of the acoustic emission science and technology and its unique applications is the goal of this article.

History

The history of acoustic emission can be divided on two main periods: pre-technological and technological. From the beginning of humankind, people were observing acoustic emission when heard cracking stones, fracturing of bones, crackling of wood in the fire and so on. Sometimes people used their experience and intentionally listened to different structures in order to detect if those are in danger. With development of work crafts, acoustic emission was helping craft makers to control quality of production. For example, during pottery making, cracking sounds were indications of fast or non-uniform pottery drying and this observation could be used for adjusting a process of pottery making.

The technological era of acoustic emission has started in the early 20th century when researches in different countries started to report about audible sounds during investigation of material deformation. So in 1916, J. Czochralski noted about a "tin cry" during twinning of tin and zinc crystals. In 1923, A. Portevin ana F. Le Chatelier reported small high frequency audible sounds that could be heard during plastic deformation of an alloy of aluminum, manganese and copper. In 1924, P. Ehrenfest and A. Yoffe observed that the process of shear deformation of salt and zinc is accompanied by clicking sounds. The first development and use of instrumentation for detection of AE was done by the seismologist F. Kishinouye during his experiments on wood in 1933. Independently, in 1936, F. Forster and E. Scheil created and applied instrumentation for registration of AE generated during martensitic transformations. In 1950, J. Kaiser investigating different engineering materials reported about the effect of the absence of acoustic emission in materials under stress levels below those previously applied on that material. This effect, bearing the name of Kaiser, is widely used in today's acoustic emission testing.

Following researches, done in recent decades have revealed that Kaiser effect is a material specific and not all materials exhibit it, for example different composites. Also, it has been shown that the effect is not observed usually in structures containing developing flaws.

In 1960s, in parallel in different countries it was proposed to use acoustic emission technology for practical non-destructive examination of different structures. So in 1961 in USA, A. Green, C. Lockman and R. Steele used acoustic emission for assessment of structural integrity of rocket motor cases fabricate for the United States Navy. In 1963, H. Dunegan, proposed to use AE for inspection of pressure vessels and in 1969 he and P. Knauss founded the first AE company in USA. In parallel, in the former Soviet Union, several scientific research institutes started investigate acoustic emission and develop its application for crack detection, material studies, non-destructive control of various structures and other, mostly military related applications. With development of the acoustic emission technology and the start of its extensive application, appeared a need in communication between researches, exchange of knowledge and common terminology. So for this purpose, starting from late 1960, there were organized acoustic emission working groups in USA, Germany, Japan and other countries. Also, there were developed standards for examination of structures, instrumentation and terminology. Today, acoustic emission testing is used practically in almost all industries and in many research centers worldwide.

The Physical Nature of the Phenomenon

Understanding the physical nature of acoustic emission in different materials is a cornerstone in the development of the acoustic emission technology. The success and the depth of the technology capabilities depend on the ability to determine the interconnection between characteristics of acoustic emission and sources it generated. However, establishing such interconnection for different materials and structures is a real scientific and technological challenge.

Material Sources of Acoustic Emission

The goals of acoustic emission examinations in industrial applications today, are detection, location and assessment of flaws in structures made of metal, concrete or composites. In these materials, fracture development in form of crack propagation is a primary source of acoustic emission. Elementary crack jumps under static or dynamic loads are followed by a rapid release of energy. A part of this energy is released in form of stress waves as a result of fast redistribution of a stress field at the crack top. The stress waves generated are elastic waves mostly but inelastic waves can be generated also when stresses exceed yield limit. This occurs, for example, at the plastic zone of a crack developing in a ductile metal.

Non-material Secondary Sources of Acoustic Emission

Acoustic emission equipment is capable of detecting and analyzing acoustic emission sources of non-material origin, for instance, mechanical sources of friction, knocks, leaks and so on. There are multiple applications in which acoustic emission technology is used for revealing leaks, machinery health monitoring, detection of dynamic stress events in structures and other using these capabilities.

Wave Propagation

Acoustic emission wave propagation out of the source it generated over the structure is always a complex mechanical puzzle. Waves of different types propagate at different velocities and with different oscillation directions. Moreover, passing through a medium, waves undergo multiple changes due to attenuation, dispersion, diffraction, scattering, reflection from boundaries, interaction with reflections and other. In those applications, where it is possible either analytically or numerically describe wave propagation, it is possible to achieve a greater accuracy in the source location and it characterization. For example, in anisotropic materials, an accurate location is possible when an effective wave velocity is incorporated in a location algorithm as function of a propagation angle.

Qualitative Types of Acoustic Emission

There are two distinct qualitative types of acoustic emission: burst and continuous. Burst is a type of emission related to individual events occurring in a material that results in discrete acoustic emission signals. Continuous is a type of emission that related to time overlapping and/or successive emission events from one or several sources that results in sustained signals. Detection, ability to distinguish and analyze signals resulting from both emission types is important for many acoustic emission applications. For example, in ductile metals most of the energy expended on fracture processes goes to development of a plastic deformation, which normally accompanied by continuous acoustic emission. This is the reason why, normally flaws at their early stages in ductile metals can be detected mostly by use of continuous emission. Also, reliable detectability of specific flaws like stress corrosion cracking and creep are depend on detection and analysis of continuous acoustic emission. At the same time, there are flaws or conditions that can be detected by burst acoustic emission, like fracture of non-metallic inclusions, breakage of corrosion products, crack jumps in brittle or at advance stages in ductile metals and other.

Acoustic Emission and Loading Conditions

Flaws are developing in materials under stress, not necessarily dynamic and/or due to exposure to different environmental conditions. Since acoustic emission is accompanying fracture processes, it is essential for the success of acoustic emission examination to learn about common flaws existing in the structure been examined and operational and stress conditions that may cause flaw origination and development. Once these factors are established, a procedure for performing AE examination can be developed. The fundamental principal of such procedure is to perform examination under the real or simulating real loading conditions that cause flaw origination and development. For example, if it is known that a thick pipe suffers from a thermal fatigue due to a large temperature gradient, it can be ineffective to examine this pipe under hydraulic pressure and ambient temperature conditions, simply because the stress distribution will be different and flaw may not develop and consequently will not actively emit acoustic emission during the test. Sometimes, it is necessary to perform a test under various operational and stress condition in order to detect and evaluate different possible types of flaws.

Application of the Acoustic Emission Method as a Diagnostic Tool for Assessment of Structural Integrity

Application of the acoustic emission as a diagnostic method, structural integrity assessment tool is possible when a qualitative or quantitative relationship between detected acoustic emission and material condition is established for a specific material and structure. There are two major approaches to achieve this goal: (1) Determining experimentally a characteristic set (fingerprints) of acoustic emission parameters and their characteristics that uniquely describe a material condition, fracture stage, flaw type and etc. For example, to find acoustic emission characteristic fingerprints of concrete cracking and rebar corrosion. (2) Establishing a theoretical relationship between acoustic emission parameters and their characteristics and material properties, fracture mechanics parameters and etc.

The Technology

Acoustic emission sensor is a device that transforms a local dynamic material displacement produced by a stress wave to an electrical signal. AE sensors are typically piezoelectric sensors with elements maid of special ceramic elements like lead zirconate titanate (PZT). These elements generate electric signals when mechanically strained. Other types of sensors include capacitive transducers, laser interferometers. Selection of a specific sensor depends on the application, type of flaws to be revealed, noise characteristics and other factors. Typical frequency range in AE applications varies between 20 kHz and 1 MHz. There are two qualitative types of sensors according to their frequency responds: resonant and wideband sensors. Thickness of piezoelectric element defines the resonance frequency of sensor. Diameter defines the area over which the sensor averages surface motion.

References

1. **Hassab J.C.,** *Under Water Signal and Data Processing*, Bocaraton, CRC Press (1989).
2. **Hackman R.G.,** *Physical Acoustics*, New York, Academic Press (1993) 22,1.
3. **Weydert M.,** *Under water Acoustics*, London, Elsevier Appl. Sci. (1992).
4. **Kupperman W.A. and Mc Donald B.E.,** *Ocean Seismo Acoustics*, New York, Plenum (1986).
5. **Sruby C.B. and Drain L.B.,** *Laser Ultrasonics*, Adam Hilger, UK (19990).
6. **Thompson D.O. and Chementi D.E.,** *Review of Progress in Quantitative Non Destructive Evaluation* (1993) 12, 495.
7. **Jones J.P.,** *Intl. J. of Imaging Systems and Tech.,* 8(1997) 61.
8. **Daft C.M.W. and Briggs G.,** *IEEE Trans. Sonics Ultrason. and Ind. Eng. Chem.*, 36 (1989) 258.
9. **Golan G. and Pitt C.W.,** *Appl. Surface Sci.*, 106 (1996) 491.
10. **Kessler L.W.,** *J. Acoust. Soc. Am.*, 55 (1974) 909.
11. **Gilmore R.S.,** *J. Phys., D: Appl. Phys.* 29(1996) 1389.
12. **Kushibiki J., Okuzawa T. and Ohashi Y.,** *J. Appl. Phys.* 87 (2000) 4395.
13. **Green leaf J.F.,** *Science*, 280 (1998) 82-85.
14. **Green leaf J.F.,** *Pro Natl. Acad. SCI*, USA, 96 (1999) 6603-6608.
15. **Fokker P.A., Dijkhuis J.I. and Dewijin H.W.,** *Phys. Rev.*, B55 (1997) 2925.
16. **Prieur J.Y., Decaud M., Joffrin J., Barre C., Stenger M. and Chevallier M.,** *Physica B.*, 219-220 (1996) 235.
17. **Makker S.S., Vasilevskiy M.I., Anda E.V., Tuyarot P.E., Webberszpil J. and Pastawski H.M.,** *J. Phys. Condens. Matter*, 10 (1998) 5905.

8

Measurement of Elastic Constants Using Ultrasonics

S. K. Shrivastava, Kailash, Rahul Gupta, Anchala and Pushpraj Singh

Introduction

A sound wave is subject to attenuation, which is dissipation of its energy content as it travels though the medium. The rate of dissipation increases with the frequency of the sound wave, and decreases with the density of the medium. This means that high frequency sound (ultrasound) has reasonable attenuation in dense solids like metals. A variety of ultrasound transducers is available for use with various materials in different applications. The ultrasound frequency, and hence the type of transducer used, depends on the attenuation experienced in a material. For measurement purposes, the trade-off is between attenuation and resolution, both of which are a function of sound frequency. Higher frequency yields proportionately higher resolution, but means higher attenuation. Therefore, for instance, one may want to use lower frequency sound when dealing with thicker, highly attenuating materials, thus compromising resolution. On the other hand, thinner, non-attenuating materials permit the use of higher frequencies and lend themselves to better resolution. Clearly, ultrasound therefore is tool that permits us to see through. opaque materials.

Stress Waves in Solids

Consider a body, which experiences a disturbance on the surface. The propagation of the disturbance through the body follows the wave equation,

$$\frac{\partial^2 \bar{u}}{\partial t^2} = c^2 \nabla^2 \bar{u} \tag{1}$$

Where $\bar{u}$ (x,y,z and t) is the displacement vector which describes the change in position of any point in the body at position (x, y, z) at time t, and where *c* is the speed of sound. In fluids such as air, sound travels as a pressure wave. In solids, which support both normal stress and shear stress, two types of waves, longitudinal and shear, may propagate. For longitudinal waves, the motion of the particles is parallel to the direction of propagation; for shear waves, the motion of particles is perpendicular to the direction of wave propagation.

It may be shown that the speed of sound of longitudinal and shear waves is given by:

$$C_L = \sqrt{\frac{E(1-v)}{\rho(1+v)(1-2v)}} \tag{2}$$

$$C_S = \sqrt{\frac{E}{2\rho(1+v)}} = \sqrt{\frac{G}{\rho}} \tag{3}$$

where 'E' is Young's modulus (or the modulus of rigidity), 'v' is Poisson's ratio and 'G' is the shear modulus. Solving these equations for E and v:

$$v = \frac{1-2\left(\frac{C_S}{C_L}\right)^2}{2-2\left(\frac{C_S}{C_L}\right)^2} \tag{4}$$

$$E = 2\rho C_S^2(1+v) \tag{5}$$

Thus given measurement of ρ, C_S and C_L, it is possible to determine E and v.

Pulse-Echo Method

The pulse-echo technique is a quick and accurate method of measuring the speed of sound in solids. In the pulse-echo ultrasonic testing technique, an ultrasound transducer generates an ultrasonic pulse and receives its echo. The ultrasonic transducer functions as both transmitter and receiver in one unit. Most ultrasonic transducer units use an electronic pulse to generate a corresponding sound pulse, using the piezoelectric effect. A short, high voltage electric pulse (less than 20 Ns in duration, 100-200 V in amplitude) excites a piezoelectric crystal, to generate an ultrasound pulse. The transducer broadcasts the ultrasonic pulse at the surface of the specimen. The ultrasonic pulse travels through the specimen and reflects off the opposite face. The transducer then listens to the reflected echoes. The ultrasound pulse keeps bouncing off the opposite faces of the specimen, attenuating with time. The time between any two echoes is the length of time required for the pulse to travel through the specimen and back to the transducer. The attenuation (amplitude decay) is exponentially with time.

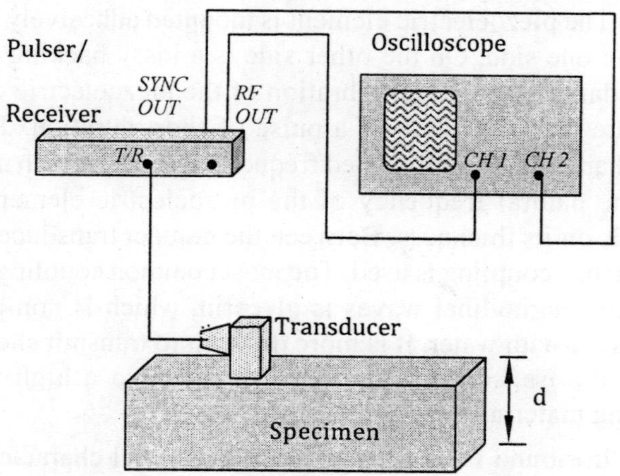

Fig. 1. Schematic Diagram of Ultrasonic Testing Apparatus.

The speed of sound in the solid can be derived from the observed round trip transit time, Δt and the measured thickness of the specimen 'd':

$$c = \frac{2d}{\Delta t} \tag{6}$$

Values for the speed of sound in a variety of solids range from 1 to 8 km/s.

Ultrasonic Transducers

In this measurement direct contact transducers are used. Such transducers are generally applicable for minimum thickness of 0.5 mm (0.020.) for metals and 0.125 mm (0.005.) for plastics. These transducers are used where accuracy requirement is not greater than ±0.025 mm (±0.001.). Figure 1 shows a schematic diagram of an ultrasonic contact transducer. The primary component is the piezoelectric quartz crystal that converts a mechanical pulse into an electrical signal, or conversely, an electrical signal to a mechanical pulse. In the pulse-echo method, the crystal functions in both modes. According to the manner in which the piezoelectric crystal is cut, it vibrates in the thickness direction, producing longitudinal waves, or in the tangential direction producing shear waves. The piezoelectric element is mounted adhesively to a wear plate on one side. On the other side is a lossy backing material, which damps the natural vibration of the piezoelectric crystal to facilitate the production of a pulse of short duration. The pulse has a characteristic bell shaped frequency spectrum with maximum near the natural frequency of the piezoelectric element, which depends on its thickness. Between the contact transducer and the specimen, a coupling is used. The most common coupling material used for longitudinal waves is glycerin, which is non-toxic and washes off with water. It is more difficult to transmit shear waves across the transducer/specimen interface, so a high viscosity coupling material is more effective.

Ultrasound is used to determine material characteristics of interest such as the presence of cracks, voids, inclusions, delaminations, porosity, part thickness, weld penetration, and braze and joint integrity. The pulse-echo technique is a quick and accurate method of measuring the speed of sound in different solids. In this technique, an ultrasonic transducer generates an ultrasonic pulse and receives its echo. The reported measurement method of the speed of sound in several solids can be used to calculate the Young's modulus and Poisson's ratio for various material.s.

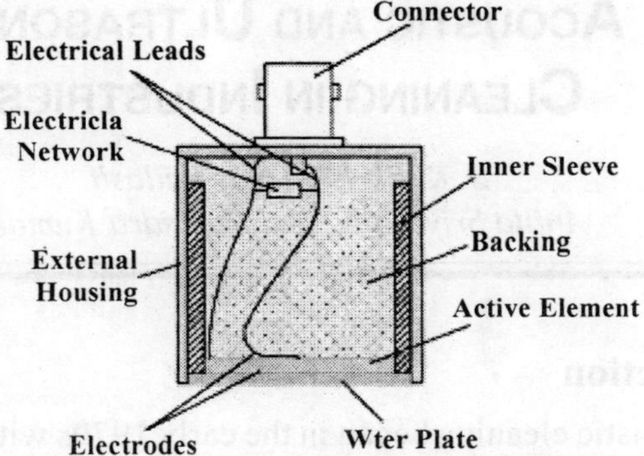

Fig. 2. Construction of Ultrasonic Transducer.

References

1. **D.S. Hughes and J.L. Kelly,** *Phys. Rev.* 92 (5), 1145-1149 (1953).
2. **G.S. Kino and al.** *J. Appl. Phys.* 50 (4), 2607-2613 (1979).
3. **R. El Guerjouma,** PhD *Thesis of Bordeaux I* (France) (1989).
4. **M. Dubuget,** PhD *Thesis of INSA* (France) 1996.
5. **J. Roux and al.** *Revue. Phys. Appl.* 20, 351-358 (1985).
6. **Krautkramer K.,** *Ultrasonic Testing of Materials*, Springer-Verlag, New York, 1969.
7. **Ensminger D.,** *Ultrasonics: Fundamentals, Technology, Applications*, M. Dekker, New York, 1988.
8. **Bolz R. E., Tuve G., L.,** *Handbook of Tables for Applied Engineering Science*, CRC Press, Florida, 1984.
9. **B. Castagnede and al.** *Ultrasonics*. Sep Vol 27 (1989).
10. **D.K. Hsu and al. J.Acoust.Soc.Am.** 92 (2), 669-675 (1992).
11. **V.A. Delgrosso and al.** *J. Acoust. Soc. Am.* 52 (5), 1442-1446 (1972).
12. **J.Y. Lechatellier.** PhD *Thesis of ENSAM* (France) 1987.

9

Acoustic and Ultrasonic Cleaning in Industries

S. K. Shrivastava, Kailash
Anjita Srivastava and Rajendra Kumar

Introduction

Acoustic cleaning began in the early 1970s with experiments using ship horns or air raid sirens. Acoustic cleaning is used wherever there is a build-up of dry materials and particulates which need to be cleaned regularly to ensure maximum efficiency, and minimize maintenance and down time. An acoustic cleaner works by generating powerful sound waves which will vibrate the dry materials differently to each other and the surrounding structures.

Ultrasonic cleaners are often used to clean jewellery, lenses and other optical parts, watches, dental and surgical instruments, fountain pens, industrial parts and electronic equipment. They are used in many jewellery workshops, watchmakers' establishments, and electronic repair workshops.

History and Design of Acoustic Cleaner

An acoustic cleaner consists of 2 parts.

- The wave generator which takes the compressed air and applies it to a diaphragm. The wave generator is usually made from solid machined stainless steel. The diaphragm within the generator is the only moving part within an acoustic cleaner and there is no danger of sparking.The diaphragm is usually manufactured from special aerospace grade titanium to ensure performance and longevity.

- The bell, which is usually made from spun 316 grade stainless steel. The bell is a resonance section or amplifier and it will tune and direct the sound waves.

An acoustic cleaner is powered by compressed air with an operating range of between 4.8 to 6.2 bars or 70 to 90 psi. The resultant sound pressure level will be around 200dB. The overall length of the acoustic cleaner will range from between 430 mm to over 3 metres long.

There are generally 4 ways to control the operation of an acoustic cleaner.

- The most common is by a simple timer.
- SCADA (Supervisory Control and Data Acquisition).
- PLC (programmable logic controller).
- Manually by Ball valve.

An acoustic cleaner will typically sound for 10 seconds and then wait for a further 500 seconds before sounding again. This ratio for on/off is approximately proportional to the working life of the diaphragm. Provided the operating environment is between – 40 and 100 °C a diaphragm should last between 3 and 5 years. The wave generator and the bell have a much longer life span and will often outlast the environment in which they operate.

The older bells which were made from cast iron were susceptible to rusting in certain environments. The new bells made from 316 spun steel have no problem with rust and are ideal for sterile environments such as found in the food industry or in pharmaceutical plants.

Operation and Performance

The majority of acoustic cleaners operate in the audio sonic range from 60 hertz up to 420 Hz. Occasionally there is a requirement to operate in the infrasonic range below 40 Hz. This would apply if there were strict noise control requirements, or there was limited plant access. There are three scientific fields which converge in the understanding of Acoustic Cleaning Technology.

- Sound propagation. This relates to an understanding of the nature of the sound waves, how they vary and how they will interact with the environment.
- Mathematics of the environment. Materials science, surface friction, distance and areas familiar to a mechanical engineer.
- Chemical engineering. The chemical properties of the powder or substance to be debonded. Especially the auto adhesive properties of the powder.

An acoustic cleaner will create a series of very rapid and powerful sound induced pressure fluctuations which are then transmitted into the solid particles of ash, dust, granules or powder. This causes them to move at differing speeds and debond from adjoining particles and the surface that they are adhering to. Once they have been separated then the material will fall off due to gravity or it will be carried away by the process gas or air stream.

The key features which determine whether or not an acoustic cleaner will be effective for any given problem are the particle size range, the moisture content and the density of the particles as well as how these characteristics will change with temperature and time.. Typically particles between 20 micrometre and 5 mm with moisture content below 8.5% are ideal. Upper temperature limits are dependent upon the melting point of the particles and acoustic cleaners have been employed at temperatures above 1000 C to remove ash build up in boiler plants.

It is important to match the operating frequencies to the requirements. Higher frequencies can be directed more accurately whilst lower frequencies will carry further, and are generally used for more demanding requirements. A typical selection of frequencies available would be as follows:

- 420 Hz for a small acoustic cleaner which might be used to clear bridging at the base of a silo.

- 350 Hz will be more powerful and this frequency can be used to unblock material build up in ID (induced draft) fans, filters, cyclones, mixers, dryers and coolers.
- 230 Hz. At this frequency the power involved is sufficient to use in most electricity generation applications.
- 75 Hz and 60 Hz. These are generally the most powerful acoustic cleaners and are often used in large vessels and silos.

Health and Safety

The introduction of acoustic cleaners has been a significant improvement in many areas of health and safety. For instance in silo cleaning - the previous solutions tended to be intrusive or destructive. Air cannons, soot blowers, external vibrators, hammering or costly man entry are all superseded by non invasive sonic horns. An acoustic cleaner requires no down time and will operate during normal usage of the site. If we take the example of silo cleaning a little further, then there are two typical problems.

Bridging

This is when the silo blocks at the outlet. Previously the problem was addressed by manual cleaning from underneath the silo which in its turn introduced significant risk from falling material when the blockage was cleared. An acoustic cleaner is able to operate from the top of a silo through in situ material to clear the blockage at the base.

Rat Holing

Compaction on the side of a silo. This not only reduces the operating volume in a silo but it also compromises quality control by disrupting the first in first out cycle. Older material compacted on the side of a silo can also start to degrade and produce dangerous gases. An acoustic cleaner will produce sound waves which will make the compacted material resonate at a different rate to the surrounding environment resulting in debonding and clearance.

Advantages of Acoustic Cleaners

- Repetitive use during operations means that there are fewer unscheduled shut downs.
- Improved material flow by the elimination of hang-ups, blocking and bridging.
- Minimisation of cross contamination by ensuring complete emptying of the environment.
- Improved cleaning and reduction of health and safety risks.
- Increased energy efficiency. Reducing the build up on heat exchange surfaces results in lower energy usage.
- Extended plant life. Aggressive cleaning regimes are avoided.
- Ease of operation. It is easy to automate the horns either at regular intervals or to tie the sounding in to changes in their environment such as pressure or flow rates.
- Importantly they prevent the material build up problem from occurring in the first place.

These advantages mean that the financial payback is often very quick. It is also possible to compare acoustic cleaners directly to alternative solutions.

- Air cannons: These are well established but are expensive with limited coverage thus requiring multi unit purchase. They are also noise intrusive and have a high compressed air consumption.
- Vibrators: These are easy to fit to an empty silo but can cause structural damage as well as contributing to powder compaction.
- Low friction linings: These are very quiet but are expensive to install. Also they are prone to erosion and can then contaminate the environment or product.
- Inflatable pads and liners: Again these are easy to install in an empty silo. They help side wall build up but have no impact on bridging. They are also hard to maintain and can cause compaction.

- Fluidization through a 1 way membrane: This can help already compacted material. However they are expensive and difficult to install and maintain. They can also contribute to mechanical interlocking and bridging.

Specific Applications For Acoustic Cleaners

- Boilers: Cleaning of the heat transfer surfaces.
- Electrostatic precipitators: Acoustic cleaners are being used for cleaning hoppers, turning vanes, distribution plates, collecting plates and electrode wires.
- Super heaters, economisers and air heaters.
- Duct work.
- Filters: Acoustic cleaners are used on reverse air, pulse jet and shaker units. They are effective in reducing pressure drop across the collection surface which will increase bag life and prevent hopper pluggage. Generally they can totally replace the both reverse air fans and shaker units and significantly reduce the compressed air requirement on pulse jet filters.
- ID fans: Acoustic cleaning helps to provide a uniform cleaning pattern even for inaccessible parts of the fan. This maintains the balance of the fan.
- Kiln inlet: Acoustic cleaners help to prevent particulate build up at the kiln inlet and this will minimize nose ring formation.
- Mechanical pre Collector: Acoustic cleaners help prevent build up around the impellers and between the tubes.
- Mills: Acoustic cleaners help maintain material flow and also prevent blockages in the pre grind silos. They also help prevent material build up in the downstream separators and fans.
- Planetary Coolers: Acoustic cleaners help prevent bridging and ensure complete evacuation.
- Precipitator: Acoustic cleaners help clean the turning vanes, distribution plates, collecting plates and electrode wires. They can either assist or replace the mechanical rapping systems.

They also prevent particulate build up in the under hoppers which would otherwise result in opacity spiking.

- Pre heaters: Used in towers, gas risers, cyclones and fans.
- Ship cargo holds: Used both to clean and de aerate current loads.
- Silos and hoppers: To prevent bridging and rat holing.
- Static cyclones: Acoustic cleaners will work both within the cyclone and with the associated duct work.

Ultrasonic Cleaning

Ultrasonic cleaner showing the removable basket in place, and a closeup of the light and timer. An ultrasonic cleaner is a cleaning device that uses ultrasound (usually from 20–400 kHz) and an appropriate cleaning solvent (sometimes ordinary tap water) to clean delicate items. The ultrasound can be used with only water but use of a solvent appropriate for the item to be cleaned and the soiling enhances the effect.

Process Characteristics

Ultrasonic cleaning uses high frequency sound waves to agitate in a liquid. Cavitation bubbles induced by the agitation act on contaminants adhering to substrates like metals, plastics, glass, rubber, and ceramics. This action also penetrates blind holes, cracks, and recesses. The intention is to thoroughly remove all traces of contamination tightly adhering or embedded onto solid surfaces. Water or other solvents can be used, depending on the type of contamination and the workpiece. Contaminants can include dust, dirt, oil, pigments, grease, polishing compounds, flux agents, fingerprints, soot wax and mold release agents, biological soil like blood, and so on. Ultrasonic cleaning can be used for a wide range of workpiece shapes, sizes and materials, and may not require the part to be disassembled prior to cleaning.

Design and Operating Principle

In an ultrasonic cleaner, the object to be cleaned is placed in a chamber containing a suitable solution (in an aqueous or organic solution, depending on the application). In aqueous cleaners, the chemical added is a surfactant which breaks down the surface tension of the water base. An ultrasound generating transducer built into the chamber, or lowered into the fluid, produces ultrasonic waves in the fluid by changing size in concert with an electrical signal oscillating at ultrasonic frequency. This creates compression waves in the liquid of the tank which 'tear' the liquid apart, leaving behind many millions of microscopic 'voids' or 'partial vacuum bubbles' (cavitation). These bubbles collapse with enormous energy; temperatures and pressures on the order of 5,000 K and 20,000 lbs per square inch are achieved; however, they are so small that they do no more than clean and remove surface dirt and contaminants. The higher the frequency, the smaller the nodes between the cavitation points, which allows for cleaning of more intricate detail. Ultrasonic transducers showing ~20 kHz and ~40 kHz stacks. The active elements (near the top) are two rings of lead zirconate titanate, which are bolted to an aluminium coupling horn. Transducers are usually made of piezoelectric material (e.g. lead zirconate titanate or barium titanate), and sometimes magnetostrictive (made of a material such as nickel or ferrite). The often harsh chemicals used as cleaners in many industries are not needed, or used in much lower concentrations, with ultrasonic agitation. Ultrasonics are used for industrial cleaning, and also used in many medical and dental techniques and industrial processes.

Cleaning Solution

Ultrasonic activity (cavitation) helps the solution to do its job; plain water would not normally be effective. The cleaning solution contains ingredients designed to make ultrasonic cleaning more effective. For example, reduction of surface tension increases cavitation levels, so the solution contains a good wetting agent (surfactant). Aqueous cleaning solutions contain detergents, wetting agents and other components, and have a large influence on the cleaning process. Correct composition of the solution is very dependent upon the item cleaned. Solutions are mostly used warm, at about 50–65 °C (122–149 °F), however, in medical applications it is generally accepted that cleaning should be at temperatures below 38 °C (100 °F) to prevent protein coagulation.

Water-based solutions are more limited in their ability to remove contaminants by chemical action alone than solvent solutions; e.g. for delicate parts covered with thick grease. The effort required to design an effective aqueous-cleaning system for a particular purpose is much greater than for a solvent system.

Some better machines (which are not unduly large) recycle the hydrocarbon cleaning fluids. Three tanks are used in a cascade. The lower tank containing dirty fluid is heated causing the fluid to evaporate. At the top of the machine there is a refrigeration coil. Fluid condenses on the coil and falls into the upper tank. The upper tank eventually overflows and clean fluid runs into the work tank where the cleaning takes place. Purchase price is higher than simpler machines, but such machines are economical in the long run. The same fluid can be reused many times, minimising wastage and pollution. Carbon tetrachloride (CCl_4, also formerly used in fire extinguishers for electrical fires) was used in the past, but is now prohibited as dangerous. If CCl_4 fumes are inhaled through a lit cigarette, carbonyl chloride ($COCl_2$, also called phosgene, a poison gas used in warfare) could be produced.

Uses

Ultrasonic baths are also used to experimentally determine the elastic constants of many anisotropic materials. Ultrasonic waves can usually only be sent through a material at right angles to the material's surface (normal incidence). In water the angle of incidence for a longitudinal wave can be set, inducing both longitudinal and transverse waves in the material. Then, by measuring the time of flight for both waves, the elastic constants can be determined.

There are various applications of acoustic cleaners and ultrasonic cleaner according to their properties and operational limitations. Most hard, non-absorbent materials (metals, plastics, etc.) not chemically attacked by the cleaning fluid are suitable for ultrasonic cleaning. Industrial ultrasonic cleaners are used in the automotive, sporting, printing, marine, medical, pharmaceutical, electroplating, disk drive components, engineering and weapons industries. While acoustic cleaners has been a significant improvement in many areas of health and safety. Acoustic cleaning is used wherever there is a build-up of dry materials and particulates which need to be cleaned regularly to ensure maximum efficiency, and minimize maintenance and down time.

References

1. **Henglein and M.J. Gutierrez.,** *J. Phys. Chem.* 97, 158, 1993
2. **Wick C. and Veilleux R.F.,** *Ultrasonic Cleaning*, Tool and Manufacturing Engineers Handbook, Materials' Finishing, and Coating, Society of Manufacturing Engineers, 3 (1985) 18.
3. **Fuchs F.J.,** *Ultrasonic Cleaning*, Metal Finishing Guidebook and Directory, Elsevier Science, (1992) 134.
4. *Ultrasonic Cleaning*, ASM Handbook, Surface Engineering, ASM International, Materials Park, 5 (1994) 44.
5. **Mackezie C Odell,** *Ultrasonic Cleaning of Interior Surfaces*, United States Patent Number 5,289,838, March 1, 1994.
6. **Ronald H Winston,** *Ultrasonic Tooth Cleaner*, United States Patent Number 5,853,290, December 29, 1998.
7. **Stan Morantz,** *Ultrasonic Golf Club Cleaning Apparatus*, United States Patent Number 5,141,009, August 25, 1992.
8. **Ronald H Winston,** *Ultrasonic Tooth Cleaner*, United States Patent Number 5,772,434, June 30, 1998.
9. **Iben Browning, Robert E. McClure,** *Ultrasonic Cleaning and Sterilizing Apparatus*, United States Patent Number 3,973,760, August 10, 1976.

10

Ultrasonic Flaw Detection in Industries

S. K. Shrivastava, Kailash Anchala and Amit Kumar

Introduction

Flaw detection is the oldest and the most common in all the applications of industrial ultrasonic testing. Since the 1940s, the laws of physics that govern the propagation of sound waves through solid materials have been used to detect hidden cracks, voids, porosity, and other internal discontinuities in metals, composites, plastics, and ceramics. High frequency sound waves reflect from flaws in predictable ways, producing distinctive echo patterns that can be displayed and recorded by portable instruments. Ultrasonic testing is completely nondestructive and safe, and it is a well established test method in many basic manufacturing, process, and service industries, especially in applications involving welds and structural metals.

Basic Theory

Sound waves are simply organized mechanical vibrations traveling through a medium, which may be a solid, a liquid, or a gas. These waves will travel through a given medium at a specific speed or velocity, in a predictable direction, and when they encounter a boundary with a different medium they will be reflected or transmitted according to simple rules. This is the principle of physics that underlies ultrasonic flaw detection.

Frequency: All sound waves oscillate at a specific frequency, or number of vibrations or cycles per second, which we experience as pitch in the familiar range of audible sound. Human hearing extends to a maximum frequency of about 20,000 cycles per second (20 KHz), while the majority of ultrasonic flaw detection applications utilize frequencies between 500,000 and 10,000,000 cycles per second (500 KHz to 10 MHz). At frequencies in the megahertz range, sound energy does not travel efficiently through air or other gasses, but it travels freely through most liquids and common engineering materials.

Velocity: The speed of a sound wave varies depending on the medium through which it is traveling, affected by the medium's density and elastic properties. Different types of sound waves (see Modes of Propagation, below) will travel at different velocities.

Wavelength: Any type of wave will have an associated wavelength, which is the distance between any two corresponding points in the wave cycle as it travels through a medium. Wavelength is related to frequency and velocity by the simple equation

$$\lambda = \frac{c}{f}$$

Where λ= wavelength, c = sound velocity, f= frequency.

Wavelength is a limiting factor that controls the amount of information that can be derived from the behavior of a wave. In ultrasonic flaw detection, the generally accepted lower limit of detection for a small flaw is one-half wavelength. Anything smaller than that will be invisible. In ultrasonic thickness gaging, the theoretical minimum measurable thickness one wavelength.

Modes of Propagation: Sound waves in solids can exist in various modes of propagation that are defined by the type of motion involved. Longitudinal waves and shear waves are the most common modes employed in ultrasonic flaw detection. Surface waves and plate waves are also used on occasion.

- A longitudinal or compressional wave is characterized by particle motion in the same direction as wave propagation, as from a piston source. Audible sound exists as longitudinal waves.

- A shear or transverse wave is characterized by particle motion perpendicular to the direction of wave propagation.
- A surface or Rayleigh wave has an elliptical particle motion and it travels across the surface of a material, penetrating to a depth of approximately one wavelength.
- A plate or Lamb wave is a complex mode of vibration in thin plates where material thickness is less than one wavelength and the wave fills the entire cross-section of the medium.

Sound waves may be converted from one form to another. Most commonly, shear waves are generated in a test material by introducing longitudinal waves at a selected angle.

Variables Limiting Transmission of Sound Waves: The distance that a wave of a given frequency and energy level will travel depends on the material through which it is traveling. As a general rule, materials that are hard and homogeneous will transmit sound waves more efficiently than those that are soft and heterogeneous or granular. Three factors govern the distance a sound wave will travel in a given medium: beam spreading, attenuation, and scattering. As the beam travels, the leading edge becomes wider, the energy associated with the wave is spread over a larger area, and eventually the energy dissipates. Attenuation is energy loss associated with sound transmission through a medium, essentially the degree to which energy is absorbed as the wave front moves forward. Scattering is random reflection of sound energy from grain boundaries and similar microstructure. As frequency goes up, beam spreading increases but the effects of attenuation and scattering are reduced. For a given application, transducer frequency should be selected to optimize these variables.

Reflection at a Boundary: When sound energy traveling through a material encounters a boundary with another material, a portion of the energy will be reflected back and a portion will be transmitted through. The amount of energy reflected, or reflection coefficient, is related to the relative acoustic impedance of the two materials. Acoustic impedance in turn is a material property defined as density multiplied by the speed of sound in a given material. For any two materials, the reflection coefficient as a percentage of incident energy pressure may be calculated through the formula

$$R = \frac{Z_2 - Z_1}{Z_2 + Z_1}$$

where R = reflection coefficient (percentage of energy reflected), Z_1 = acoustic impedance of first material, Z_2 = acoustic impedance of second material

For the metal/air boundaries commonly seen in ultrasonic flaw detection applications, the reflection coefficient approaches 100%. Virtually all of the sound energy is reflected from a crack or other discontinuity in the path of the wave. This is the fundamental principle that makes ultrasonic flaw detection possible.

Angle of Reflection and Refraction: Sound energy at ultrasonic frequencies is highly directional and the sound beams used for flaw detection are well defined. In situations where sound reflects off a boundary, the angle of reflection equals the angle of incidence. A sound beam that hits a surface at perpendicular incidence will reflect straight back. A sound beam that hits a surface at an angle will reflect forward at the same angle.

Sound energy that is transmitted from one material to another bends in accordance with Snell's Law of refraction. Again, a beam that is traveling straight will continue in a straight direction, but a beam that strikes a boundary at an angle will be bent according to the formula:

$$\frac{Sin\theta_1}{Sin\theta_2} = \frac{V_1}{V_2}$$

where θ_1 = incident angle in first material, θ_2 = refracted angle in second material, V_1 = sound velocity in first material, V_2 = sound velocity in second material

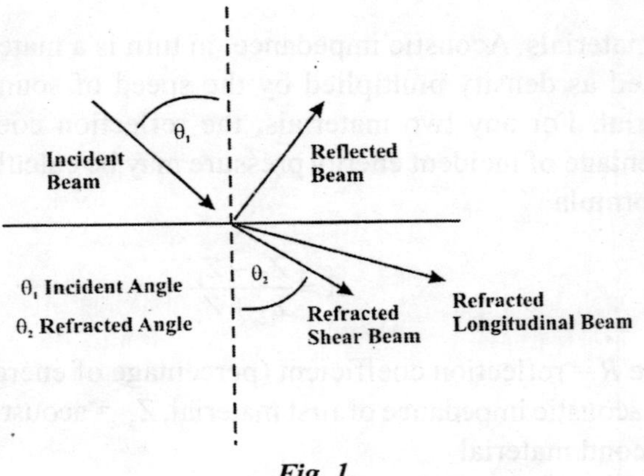

Fig. 1.

Ultrasonic Transducers

In the broadest sense, a transducer is a device that converts energy from one form to another. Ultrasonic transducers convert electrical energy into high frequency sound energy and vice versa. In this measurement direct contact transducers are used. Such transducers are generally applicable for minimum thickness of 0.5 mm (0.020) for metals and 0.125 mm (0.005) for plastics. These transducers are used where accuracy requirement is not greater than ±0.025 mm (±0.001). Fig. 2 shows a schematic diagram of an ultrasonic contact transducer. The primary component is the piezoelectric quartz crystal that converts a mechanical pulse into an electrical signal, or conversely, an electrical signal to a mechanical pulse. In the pulse-echo method, the crystal functions in both modes. According to the manner in which the piezoelectric crystal is cut, it vibrates in the thickness direction, producing longitudinal waves, or in the tangential direction producing shear waves. The piezoelectric element is mounted adhesively to a wear plate on one side. On the other side is a lossy backing material, which damps the natural vibration of the piezoelectric crystal to facilitate the production of a pulse of short duration. The pulse has a characteristic bell shaped frequency spectrum with maximum near the natural frequency of the piezoelectric element, which depends on its thickness. Between the contact transducer and the specimen, a coupling is used. The most common coupling material used for longitudinal waves is glycerin, which is non-toxic and washes off with water. It is more difficult to transmit shear waves across the transducer/specimen interface, so a high viscosity coupling material is more effective.

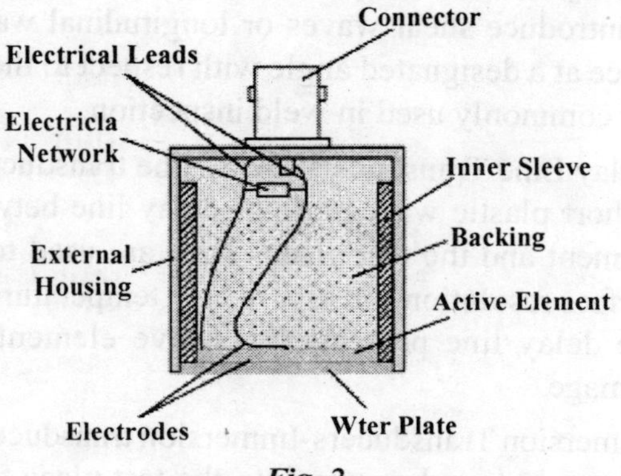

Fig. 2.

Cross Section of Typical Contact Transducer

Typical transducers for ultrasonic flaw detection utilize an active element made of a piezoelectric ceramic, composite, or polymer. When this element is excited by a high voltage electrical pulse, it vibrates across a specific spectrum of frequencies and generates a burst of sound waves. When it is vibrated by an incoming sound wave, it generates an electrical pulse. The front surface of the element is usually covered by a wear plate that protects it from damage, and the back surface is bonded to backing material that mechanically dampens vibrations once the sound generation process is complete. Because sound energy at ultrasonic frequencies does not travel efficiently through gasses, a thin layer of coupling liquid or gel is normally used between the transducer and the test piece. There are five types of ultrasonic transducers commonly used in flaw detection applications:

- Contact Transducers-As the name implies, contact transducers are used in direct contact with the test piece. They introduce sound energy perpendicular to the surface, and are typically used for locating voids, porosity, and cracks or delaminations parallel to the outside surface of a part, as well as for measuring thickness.

- Angle Beam Transducers-Angle beam transducers are used in conjunction with plastic or epoxy wedges (angle beams) to introduce shear waves or longitudinal waves into a test piece at a designated angle with respect to the surface. They are commonly used in weld inspection.
- Delay Line Transducers-Delay line transducers incorporate a short plastic waveguide or delay line between the active element and the test piece. They are used to improve near surface resolution and also in high temperature testing, where the delay line protects the active element from thermal damage.
- Immersion Transducers-Immersion transducers are designed to couple sound energy into the test piece through a water column or water bath. They are used in automated scanning applications and also in situations where a sharply focused beam is needed to improve flaw resolution.
- Dual Element Transducers — Dual element transducers utilize separate transmitter and receiver elements in a single assembly. They are often used in applications involving rough surfaces, coarse grained materials, detection of pitting or porosity, and they offer good high temperature tolerance as well.

Ultrasonic Flaw Detector

Modern ultrasonic flaw detectors are small, portable, microprocessor-based instruments suitable for both shop and field use. They generate and display an ultrasonic waveform that is interpreted by a trained operator, often with the aid of analysis software, to locate and categorize flaws in test pieces. They will typically include an ultrasonic pulser/receiver, hardware and software for signal capture and analysis, a waveform display, and a data logging module. While some analog-based flaw detectors are still manufactured, most contemporary instruments use digital signal processing for improved stability and precision.

Fig. 3.

The pulser/receiver section is the ultrasonic front end of the flaw detector. It provides an excitation pulse to drive the transducer, and amplification and filtering for the returning echoes. Pulse amplitude, shape, and damping can be controlled to optimize transducer performance, and receiver gain and bandwidth can be adjusted to optimize signal-to-noise ratios.

Modern flaw detectors typically capture a waveform digitally and then perform various measurement and analysis function on it. A clock or timer will be used to synchronize transducer pulses and provide distance calibration. Signal processing may be as simple as generation of a waveform display that shows signal amplitude versus time on a calibrated scale, or as complex as sophisticated digital processing algorithms that incorporate distance/amplitude correction and trigonometric calculations for angled sound paths. Alarm gates are often employed to monitor signal levels at selected points in the wave train to flag echoes from flaws.

The display may be a CRT, a liquid crystal, or an electro luminescent display. The screen will typically be calibrated in units of depth or distance. Multicolor displays can be used to provide interpretive assistance.

Internal data loggers can be used to record full waveform and setup information associated with each test, if required for documentation purposes, or selected information like echo amplitude, depth or distance readings, or presence or absence of alarm conditions.

Procedure

Ultrasonic flaw detection is basically a comparative technique. Using appropriate reference standards along with a knowledge of sound wave propagation and generally accepted test procedures, a trained operator identifies specific echo patterns corresponding to the echo response from good parts and from representative flaws. The echo pattern from an test piece may then be compared to the patterns from these calibration standards to determine its condition.

Straight Beam Testing: Straight beam testing utilizing contact, delay line, dual element, or immersion transducers is generally employed to find cracks or delaminations parallel to the surface of the test piece, as well as voids and porosity. It utilizes the basic principle that sound energy traveling through a medium will continue to propagate until it either disperses or reflects off a boundary with another material, such as the air surrounding a far wall or found inside a crack. In this type of test, the operator couples the transducer to the test piece and locates the echo returning from the far wall of the test piece, and then looks for any echoes that arrive ahead of that backwall echo, discounting grain scatter noise if present. An acoustically significant echo that precedes the backwall echo implies the presence of a laminar crack or void. Through further analysis, the depth, size, and shape of the structure producing the reflection can be determined. In some specialized cases, testing is performed in a through transmission mode, where sound energy travels between two transducers placed on opposite sides of the test piece. If a large flaw is present in the sound path, the beam will be obstructed and the sound pulse will not reach the receiver.

Angle Beam Testing: Cracks or other discontinuities perpendicular to the surface of a test piece, or tilted with respect to that surface, are usually invisible with straight beam test techniques because of their orientation with respect to the sound beam. Such defects can occur in welds, in structural metal parts, and many other critical components. To find them, angle beam techniques are used, employing either common angle beam (wedge) transducer assemblies or immersion transducers aligned so as to direct sound energy into the test piece at a selected angle. The use of angle beam testing is especially common in weld inspection.

Typical angle beam assemblies make use of mode conversion and Snell's Law to generate a shear wave at a selected angle (most commonly 30, 45, 60, or 70 degrees) in the test piece. As the angle of an incident longitudinal wave with respect to a surface increases, an increasing portion of the sound energy is converted to a shear wave in the second material, and if the angle is high enough, all of the energy in the second material will be in the form of shear waves. There are two advantages to designing common angle beams to take advantage of this mode conversion phenomenon. First, energy transfer is more efficient at the incident angles that generate shear waves in steel and similar materials. Second, minimum flaw size resolution is improved through the use of shear waves, since at a given frequency, the wavelength of a shear wave is approximately 60% the wavelength of a comparable longitudinal wave.

Typical Angle Beam Assembly: The angled sound beam is highly sensitive to cracks perpendicular to the far surface of the test piece (first leg test) or, after bouncing off the far side, to cracks perpendicular to the coupling surface (second leg test). A variety of specific beam angles and probe positions are used to accommodate different part geometries and flaw types, and these are described in detail in appropriate inspection codes and procedures such as ASTM E-164 and the AWS Structural Welding Code.

Ultrasonic flaw detection is a technology that is widely used for locating and sizing hidden cracks, voids, disbonds, and similar discontinuities in welds, forgings, billets, axles, shafts, tanks and pressure vessels, turbines, structural components, as well as in composites and fiberglass. Modern digital flaw detectors offer unmatched capability and versatility for a wide variety of test applications.

References

1. **Birks A.S., Green R.E. and McIntire P.,** *The Nondestructive Testing Handbook,* Second Edition. Vol.7, American Society for Nondestructive Testing (1991) 893.
2. Nondestructive Evaluation and Quality Control. *ASM Handbook.* Vol.17, ASM International, USA (1994) 795.
3. **Baldev Raj, T. Jayakumar, B.P.C. Rao,** *Indian Academy Proc. in Engineering Sciences,* Vol. 20, 1995, pp. 5-38.
4. **B.P.C. Rao,** *Encyclopedia of Materials: Science and Technology,* Elsevier Science Ltd, September 2001, pp 6043-6046.
5. **B. Venkatraman, V. Manoharan, T. Jayakumar and P. Kalyanasundaram,** *Proc. of 15th World Conference on NDT,* Rome, Oct. 2000.
6. **B. Venkatraman, S. Saravanan, T. Jayakumar, P. Kalyanasundaram and Baldev Raj,** *Proc. of 14th World Conf. on Non Destructive Testing* (14th WCNDT), New Delhi, India, 1996, Vol.3, pp. 1401-1404.
7. **T. Jayakumar and K.V. Rajkumar,** *J. Pure and Applied Ultrasonics,* 22, 2000, pp. 89-105.
8. **B. P. C. Rao, T. Jayakumar, P. Kalyanasundaram and Baldev Raj,** *J. of NDE* (India), Vol.19, No.2, June 1999, pp 23-28.
9. **M.Thavasimuthu, K.V. Rajkumar, T. Jayakumar, P. Kalyanasundaram and Baldev Raj,** *Plenum Publ. Corp.,* New York, Vol. 18B, pp. 1987-1993.
10. **B.P.C. Rao, Baldev Raj, T. Jayakumar & P. Kalyanasundaram,** *NDT&E International,* Vol. 35, No.6, 2002, pp 393-398.
11. **B.P.C. Rao, Baldev Raj, T. Jayakumar & P. Kalyanasundaram,** *Nondestr. Test. Eval.,* Vol. 17, 2000, pp 41-57.
12. **Lynnworth, Lawrence C.,** *Ultrasonic Measurements for Process Control,* Academic Press (1989).

11

How to Grow Great Crystals

S. K. Verma, S. K. Shrivastava
and Anjita Shrivastava

Do you want to grow great crystals? These are general instructions for growing crystals that you can use for most crystal recipes. You can find the recipes and information on crystal structures in the Growing Crystals section. Here are the basics, to get you started and help you troubleshoot problems.

What Are Crystals?

Crystals are structures that are formed from a regular repeated pattern of connected atoms or molecules. Crystals grow by a process termed *nucleation*. During nucleation, the atoms or molecules that will crystallize (solute) are dissolved into their individual units in a solvent. The solute particles contact each other and connect with each other. This subunit is larger than an individual particle, so more particles will contact and connect with it. Eventually, this crystal nucleus becomes large enough that it falls out of solution (crystallizes). Other solute molecules will continue to attach to the surface of the crystal, causing it to grow until a balance or equilibrium is reached between the solute molecules in the crystal and those that remain in the solution.

The Basic Technique

- Make a saturated solution.
- Start a garden or grow a seed crystal.
- Continue growth.

In order to grow a crystal, you need to make a solution which maximizes the chances for the solute particles to come together and form a nucleus, which will grow into your crystal. This means you will want a concentrated solution with as much solute as you can dissolve (saturated solution). Sometimes nucleation can occur simply through the interactions between the solute particles in the solution (called unassisted nucleation), but sometimes it's better to provided a sort of meeting place for solute particles to aggregate (assisted nucleation). A rough surface tends to be more attractive for nucleation than a smooth surface. As an example, a crystal is more likely to start forming on a rough piece of string than on the smooth side of a glass.

Make a Saturated Solution

It's best to start your crystals with a saturated solution. A more dilute solution will become saturated as the air evaporates some liquid, but evaporation takes time (days, weeks). You will get your crystals more quickly if the solution is saturated to begin with. Also, there may come a time when you need to add more liquid to your crystal solution. If your solution is anything but saturated, then it will undo your work and actually dissolve your crystals! Make a saturated solution by adding your crystal solute (e.g., alum, sugar, salt) to the solvent (usually water, although some recipes may call for other solvents). Stirring the mix will help to dissolve the solute. Sometimes you may want to apply heat to help the solute dissolve. You can use boiling water or sometimes even heat the solution on the stove, over a burner, or in a microwave.

Growing a Crystal Garden

If you just want to grow a mass of crystals or a crystal garden, you can pour your saturated solution over a substrate (rocks, brick, sponge), cover the setup with a paper towel or coffee filter to keep out dust, and allow the liquid to slowly evaporate.

Growing a Seed Crystal

On the other hand, if you are trying to grow a larger single circunl, you will need to obtain a seed crystal. One method of getting a seed crystal is to pour a small amount of your saturated solution onto a plate, let the drop evaporate, and scrape the crystals formed on the bottom to use as seeds. Another method is to pour saturated solution into a very smooth container (like a glass jar) and dangle a rough object (like a piece of string) into the liquid. Small crystals will start to grow on the string, which can be used as seed crystals.

Crystal Growth and Housekeeping

If your seed crystal is on a string, pour the liquid into a clean container (otherwise crystals will eventually grow on the glass and compete with *your* crystal), suspend the string in the liquid, cover the container with a paper towel or coffee filter (don't seal it with a lid!), and continue to grow your crystal. Pour the liquid into a clean container whenever you see crystals growing on the container.

If you selected a seed from a plate, tie it onto a *nylon* fishing line (too smooth to be attractive to crystals, so your seed can grow without competition), suspend the crystal in a clean container with saturated solution, and grow your crystal the same way as with seeds that were originally on a string.

Keeping Your Treasures

Crystals that were made from a water (aqueous) solution will dissolve somewhat in humid air. Keep your crystal beautiful by storing it in a dry, closed container. You may wish to wrap it in paper to keep it dry and prevent dust from accumulating on it. Certain crystals can be protected by being sealed with an acrylic coating (like Future floor polish), although applying the acrylic will dissolve the outermost layer of the crystal.

12

ELECTRONIC STRUCTURE OF DISORDERED ALLOYS

S. K. Verma and Man Singh

Introduction

There are a rich variety of disordered alloys having diverse electrical, magnetic, transport, optical and super conducting properties in nature; these properties make them very useful for technological applications. For understanding these properties knowledge of their electronic structure is essential. In disordered alloys due to the lackness of translational invariance, which is a characteristic of ordered solids, Bloch's theorem, which greatly simplifies the electronic structure calculation of ordered solids, is inapplicable to these systems. In other words, the standard band theory method developed for ordered solids cannot be applied to disordered alloys.

The main subject is the calculations of electronic structure of substitutional random binary alloys which are the simplest kind of disordered systems. In such alloys there exists a lattice structure but each lattice point can be occupied by either of the constituent atoms. Examples of such systems are a-brass Cu-Zn, Cu-Ge, Cu-Pd, Cu-Ni, Cu-Au, Ag-Pd, Nb-Mo, Cr-V and Cr-Mo etc. Such an alloy will be denoted by $A_X B_Y$, Where X and Y are concentrations of A and B types of atoms in the alloy, respectively.

A great progress has been made in understanding the electronic structure of the disordered alloys during last three decades, by the application of Coherent Potential Approximation (CPA). The CPA is a mean field approximation, which for the sake of configurational averaging replaces a disordered alloy ($A_X B_Y$) by an ordered solid of effective atoms. These effective atoms are determined by using the self-consistent condition that if an A (B) atom is embedded in this effective medium, the average scattering with respect to the medium is zero i.e.

$$x.t_A^{CPA} + y.t_B^{CPA} = 0 \tag{1}$$

where t_A^{CPA} is atomic scattering operator for an A or B atom embedded in the effective medium. A simpler approximation is average t-matrix approximation (ATA) in which the t-matrix corresponding to the effective atom is

$$t^{ATA} = x.t_A + y.t_B \tag{2}$$

where $t_{A(B)}$ is atomic scattering operator for an isolated $A(B)$ atom. Another simple approximation is virtual crystal approximation (VCA) in which the potential corresponding to the effective atoms is assumed to be

$$V^{VCA} = x.V_A + y.V_B \tag{3}$$

where $V_{A(B)}$ is the potential of $A(B)$ atom. This approximation is good only when the difference between the two constituents potential is small. Note that all these approximations (CPA, ATA, VCA) are single-site approximations; i.e. they neglect correlated scattering from clusters of atoms. The CPA has been found to be the best single-site approximation for calculation of electronic structure of disordered alloys.

A frequently quoted model in alloy theory is the rigid band model. In this model the potentials of the constituent atoms are assumed to be the same. However, Fermi energy (E_F) is adjusted to give the required number of valence electrons (N^V) per atom in the alloy as

$$N^V = \int_{-\infty}^{E_F} \rho(E) dE \tag{4}$$

where $\rho(E)$ is the density of states (DOS) per atom of the host system. The rigid band model is a crude approximation and does not explain many experimental results. Because of its simplicity, it is frequently used as a first step to get an idea of the electronic structure of the alloy.

There is very large number of interacting electrons ($\approx 10^{23}$) interacting with nuclear potentials in a solid. Therefore, the calculation of energy levels of such a system is a many-body problem, which cannot be solved exactly. However, this problem can be solved to a great degree of accuracy by using the one electron approximation. In this approximation, the many-body problem is reduced to a problem of one electron moving in some effective potential which is determined self-consistently. Muffin-tin approximation is further used to simplify the calculation of this potential. In this approximation the potential is assumed to be spherically symmetric within a sphere and Wigner-Seitz cell as

$$V(\vec{r}) = \begin{cases} V_{A(B)}(r) & , \ for\ r = < r_m \\ V^I_{A(B)} & , \ for\ r > r_m \end{cases} \qquad (5)$$

The V^I is a constant interstitial potential as

$$V^I_{A(B)} = 3 \int_{r_m}^{r_{ws}} V(r)\, r^2\, dr \Big/ (r_{ws}^3 - r_m^3) \qquad (6)$$

where $V(r)$ is a spherically symmetric potential, which, it is hoped, will be slowly varying in the region between the muffin-tin sphere and the Wigner-Seitz sphere of radius r_{ws} For a pure $A(B)$ solid. $V^I_{A(B)}$ defines muffin-tin zero, i.e. this constant is subtracted from the potential (5) making interstitial potential zero. For the alloy, this is calculated as

$$V^I_{ALLOY} = x.V^I_A + y.V^I_B \qquad (7)$$

Note that muffin-tin potentials are non-overlapping. The muffintin approximation is a reasonably good approximation for the potential as can be seen by referring to the book by Moruzzi et al. Using this approximation, they have calculated various properties such as cohesive energy, bulk modulus, density of states and Fermi energy. They got very good agreements with experimental results and were able to explain general trends. Recent experience with alloys show that the muffin-tin approximation is also a good approximation for metallic alloys. The muffin-tin potentials for the constituent atoms are constructed by using the local density approximation (LDA) of the density functional theory. The potential can be written as

$$V_{A(B)}(\vec{r}) = V_{COUL}^{A(B)}(\vec{r}) + V_{XC}^{A(B)}(\vec{r}) \tag{8}$$

where coulomb, nuclear and exchange-correlation contributions are

$$V_{COUL}^{A(B)}(\vec{r}) = \int_{\Omega} d\vec{r}\, \frac{P_{A(B)}(\vec{r}\,')}{|\vec{r} - \vec{r}\,'|} \tag{9}$$

$$V_{NUC}^{A(B)}(\vec{r}) = -e^2 Z_{A(B)} \Big/ r \tag{10}$$

$$V_{XC}^{A(B)}(\vec{r}) = \frac{\delta E_{XC}\left[P_{A(B)}(\vec{r})\right]}{\delta P_{A(B)}(\vec{r})} \tag{11}$$

Where $P_{A(B)}$ is the electron charge density in $A(B)$ cell, Z is the atomic number and Ω denotes the integral over the Wigner-Seitz unit cell. Here, E_{XC} is the exchange-correlation energy functional, which in the local density approximation is given by

$$E_{XC}\{p\} \approx \int P(r)\varepsilon_{XC}(P(\vec{r}))d\vec{r} \tag{12}$$

Where $\varepsilon_{XC}(\rho)$ is the contribution of exchange and correlation to the total energy (per electron) in a homogeneous but interacting electron gas of density ρ For paramagnetic case, von Barth-Hedin form of ε_{XC} is

$$\varepsilon_{XC}^{B}(r_S) = -\frac{0.91633}{r_S} - 0.045F\left(r_S/_{21}\right) \tag{13}$$

where r_s is given by

$$\frac{4\pi}{3}(r_S)^3 = \left(\frac{1}{\rho}\right) \tag{14}$$

and function

$$F(x) = (1 + x^3) \text{In} (1 + 1/x) - x^2 + x/2 - 1/3 \tag{15}$$

Gunnarsson and Lundqvist form of ε_{XC} for paramagnetic case is[17]

$$\varepsilon_{XC}^{G}(r_S) = -\frac{0.458}{r_S} - 0.0333F\left(\frac{r_S}{11.4}\right) \tag{16}$$

vosko et.al. form of ε_{XC} for paramagnetic case is

$$\varepsilon_{XC}^{V}(r_S) = \frac{-0.91633}{r_S} + A\left[In\left[\frac{x^2}{X(x)}\right] + \frac{2b}{Q}\tan^{-1}\left[\frac{Q}{2x+b}\right]\right.$$

$$\left. - \frac{bx_0}{X(x_0)}\left[In\left[\frac{(x-x_0)^2}{X(x)}\right] + \frac{2(b+2x_0)}{Q}\tan^{-1}\left[\frac{Q}{2x+b}\right]\right]\right]$$

Where

$$X(x) = x^2 + bx + c, Q = (4c - b^2)^{1/2} \text{ and } x = (r_s)^{1/2} \tag{17}$$

Here, $A = 0.0621814$, $x_0 = -0.10498$, $b = 3.72744$ and $c = 12.9352$

It has been seen that use of different exchange-correlation does not give large differences in the calculation of various properties. For example differences in cohesive energies in Li, Na, K and Rb were found to be less than 8 mRy. Because Moruzzi et al, have found good agreement between theory and experiment for several metals including Cr and Al, using von Barth-Hedin form of the exchange-correlation potential.

The CPA theory has been very successful as a single-site approximation to calculate the electronic structure of random substitutional disordered alloys during last thirty years. This theory when applied to muffin-tin model of disordered alloys, is known as Korringa-Kohn-Rostoker–CPA (KKR-CPA) theory. In the perfect crystal limit, the KKR-CPA theory reduced to the standard KKR band theory. In earlier stage of application, potentials V_A and V_B were not determined self-consistently. In the most sophisticated application of the KKR-CPA, within the framework of the local density approximation V_A and V_B are determined self-consistently. This fully charge-self-consistent KKR-CPA theory is a first principle parameter-free theory of the electronic structure of random alloys.

References

1. **H. Ehrenreich and L.M. Schwartz,** *Solid State Physics* 31, (1976).
2. J.S. Faulkner in Progress in Materials Science, ed. By T. Massalski (Pergamon, New York, 1982) vol. 27; J.S. Faulkner and G.M. Stocks, Phys. Rev. B21, 3222 (1980).
3. **A. Bansil,** *Electronic Band Structure and its Applications*, ed. M. Yussouff, (Springer-Verlag, Berlin, 1987), p 273; R.Prasad, Method of Electronic Structure Calculations, ed. O.R. Anderson, V. Kumar and A. Mookerjee (Singapore: World-Scientific) p 211; Indian J. pure Appl. Phys. 29, 255 (1991).
4. **L. Nordhem,** *Ann. Phys.* (Leipzig) 9, 607 and 641 (1931); F. Bassani and D. Brust, Phys. Rev. 131, 1524 (1963); H. Amar, K.H. Johnson, and C.B. Sommer's, Phys. Rev., 153 655 (1967); M.M. Pant and S. K. Joshi, Phys. Rev., 184, 635 (1969).
5. **N.F. Mott and H. Jones,** *The Theory of the Properties of Metals and Alloys* (Clarendon, Oxford, 1936); J. Friedel, Nuovo cimento suppl. 7, 287 (1958).
6. **T.L. Loucks,** *Augmented Plane Wave Method*, (Benjamin, New York, 1967).
7. **V.L. Moruzzi, J.F. Janak and A.R. Williams,** *Calculated Electronic Properties of Metals* (Pergamon, New York, 1978).
8. **A. Bansil, L.M. Schwartz and H. Ehrenreich,** *Phys. Rev.* B12, 2893 (1975).
9. **A. Bansil, H. Ehrenreich, L.M. Schwartz and R.E. Watson,** *Phys. Rev.* B9, 445 (1974).

10. **H. Asonen, M. Lindroos, M. Pessa, R. Prasad, R.S. Rao and A. Bansil,** *Phys. Rev.* B25, 7075 (1982).

11. **R. Prasad and A. Bansil,** *Phys. Rev. Letters*, 48, 113 (1982).

12. **R.S. Rao, A. Bansil, H. Asinen and M. Pessa,** *Phys. Rev.* B29, 1713 (1984).

13. **R. Prasad, S. C. Papadopoulos and A. Bansil, Phys. Rev.** B23 2607 (1981).

14. **R. Prasad, R. S. Rao and A. Bansil,** *Excitations in Disordered System*, ed. M. P. Thorpe (Plenum, New York, 1982).

15. **P. Hohenberg and W. Kohn,** *Phys. Rev.* 136, B 864 (1964). S. Lundqvist and N.H. March, Theory of the Inhomogeneous Electron Gas, (Plenum, New York, 1983). ; J. Callaway and N. H. March, Solid State Physics, 38, 135 (1984).

16. **U. Von Barth and L. Hedin,** *J. Phys. C: Solid state Phys.* 5, 1629 (1972).

17. **O. Gunnarsson and B. L. Lundqvist,** *Phys. Rev.* B13, 4274 (1976).

18. **S. H. Vosko and L. Wilk,** *Phys. Rev*. B 22, 3812 (1980); S. H. Vosko, L. Wilk and M. Nusair, Can. J. Phys. 58, 1200 (1980).

13

KKR-CPA Theory

S. K. Verma and
Man Singh

Introduction

To understand the electronic structure of disordered alloys, the Korringa-Kohn-Rostoker Coherent Potential Approximation (KKR-CPA) theory has been very successful in last three decades. It is basically an application of the coherent potential approximation (CPA)[1-3] to muffin-tin potential of the disordered alloys. In the CPA, one replaces a disordered alloy by an ordered array of effective atoms which are determined by a self-consistent condition. The muffin-tin approximation of the potential plays an important role in simplifying the theory. Because the constituent muffin-tin potentials are spherically symmetric within a certain radius r_m, various quantities such as t-matrices can be expressed in simple forms in the angular momentum space. The non-overlapping nature of the constituent potentials further simplifies the multiple scattering equations to tractable forms.

The KKR-CPA reduced to standard KKR band theory in the perfect crystal limit. Within the local density approximation of the density-functional theory (DFT)[15-18], it can be made fully charge self-consistent. Thus charge self-consistent KKR-CPA provides a first principles parameter-free theory of electronic structure of disordered alloys. The theory rests on the same theoretical footing as the band theory of pure metals.

Korring A-Kohn-Rostoker Coherent-Potential-Approximation (KKR-CPA)

Consider a disordered substitutional binary alloy A_XB_Y of A and B atoms with concentrations X and Y respectively. The one electron potential in the alloy can be written as

$$V(\vec{r}) = \sum_{i=1}^{N} V a(\vec{r}_i) \tag{1}$$

where the vectors $\vec{r}_i$ are defined by $\vec{r}_i = \vec{r} - \vec{R}_i, \vec{R}_i$ are the location of the lattice points and N is the total number of sites in the solid. $V_\alpha(\vec{r})$ (a=A,B) is the potential due to A or B atom and is assumed to be the muffin-tin form and does not overlap with each other;

$$V_\alpha(\vec{r}_i) = V\alpha(r_i), \ r_i < r_m$$

$$V_\alpha(\vec{r}_i) = \text{Constant}, r_i > r_m \tag{2}$$

where r_m is the muffin-tin radius.

The Hamiltonian for a certain configuration of the alloy in atomic mass unit is

$$H = -\nabla^2 + V(\vec{r}) \tag{3}$$

The Green's function in operator notation for a system of scatterers is defined as

$$G = (EI - H)^{-1} \tag{4}$$

where E and I are energy and identity matrix. The equation (4) can be expanded using the Dyson's equation as

$$G = G_0 + G_0 TG_0 \tag{5}$$

where G_0 is the free electron Green's function and

$$\mathbf{T} = V + VG_0\mathbf{T} \tag{6}$$

is the total scattering operator. T may be written as

$$\mathbf{T} = \sum_{i,j}^{N} T_{i,J} \tag{7}$$

where the operators $T_{i,j}$ are called path operators and satisfy the following multiple scattering equation

$$T_{i,j} = t_i \delta_{i,j} + t_i G_0 \sum_{k \neq i} T_{kj} \tag{8}$$

Here,

$$t_i = V_i (1 + G_0 t_i) \tag{9}$$

is the t-matrix that describes the scattering from an isolated potential on the i-th site. Equation (9) in angular momentum space has on-the-energy-shell matrix elements as

$$t^{l}_{A(B)}(\chi) = -(\chi)^{-1} \exp(i\delta_l)\sin(\delta_l) \tag{10}$$

where ***l*** is the angular momentum index, $\chi = (E)^{1/2}$ and δ_l are phase shifts. The free electron Green's function G_0 in real space can be expressed as

$$G_0(\vec{r}, \vec{r}\,') = -i.\chi \sum_{L} J_L(\chi.r_<) h_L(\chi.r_>) Y_L(\hat{r}_n) Y_L(\hat{r}_m) \delta_{nm}$$

$$+ \sum_{LL'} Y_L(\hat{r}_n) J_L(\chi.r_n) B^{LL'}_{nm} J_{L'}(\chi.r'_m) Y_{L'}(\hat{r}'_m) \tag{11}$$

Where $\vec{r} = \vec{r}_n + \vec{R}_n$, $\vec{r}\,' = \vec{r}\,'_m + \vec{R}_m$ and $r_<$ ($r_>$) is the smaller (greater) of the variable r_n and r'_m. Also L= (l,m) is a composite index, $Y_L(\hat{x})$ is the real spherical harmonic and J_L and h_L are the spherical Bessel and Hankel functions respectively. The matrix B is defined as

$$B^{LL'}_{nm}(\chi) = -\left[4\pi i\, \chi \sum_{L_1} i^{L-L'-L_1} C_{LL'L_1} Y_{L_1}(\vec{R}_m - \vec{R}_n) * h^{+}_{L_1}\left(\chi |\vec{R}_n - \vec{R}_m|\right) \right] (1 - \delta_{nm}) \tag{12}$$

where

$$C_{LL'L_1} = \int d\Omega_X Y_{L'}(\hat{x}) Y_{L_1}(\hat{x}) \tag{13}$$

and h^{+}_{L} denotes the outgoing Hankel function.

Starting from equation (8) and after some algebraic manipulations, on the energy-shell matrix elements of the path operator can finally be expressed as

$$T_{nm}^{LL'}(\chi) = \left[\left\{t^{-1}(\chi) - B(\chi)\right\}^{-1}\right]_{nm}^{LL'} \qquad (14)$$

The Fourier transform of equation (12)

$$\left[B_{\vec{k}}(\chi)\right]_{LL'} = (1/N)\sum_{nm} \exp\left[-i\vec{k}(\vec{R}_n - \vec{R}_m)\right] B_{nm}^{LL'}(\chi) \qquad (15)$$

is related to the well-known KKR structure functions. Equations (10) and (15) are very important in the KKR theory; $t(\chi)$ depends only on the potential while $B_{\vec{k}}(\chi)$ depends only on the lattice structure. These two matrices play a central role in the KKR-CPA theory.

Now the Green's function G can be written as

$$G(E,\vec{r},\vec{r}') = \sum_{LL'} Z_L^n(E,\vec{r}_n) T_{nm}^{LL'}(\chi) Z_{L'}^m(E,\vec{r}_m) - \delta_{nm} \sum_{L} Z_L^n(E,\vec{r}_n) J_L^n(E,\vec{r}_n) \qquad (16)$$

where $\vec{r}$ and $\vec{r}\,'$ are within n[th] and m[th] muffin-tin spheres and the wave functions $Z_L^{n(m)}(E,\vec{r}_{n(m)})$ and $J_L^{n(m)}(E,\vec{r}_{n(m)})$ are respectively the regular and irregular solutions of the differential equation.

$$\left[\ -\nabla^2 + V_{n(m)}(\vec{r}_{n(m)}) - E\ \right] Z_L^{n(m)}(E,\vec{r}_{n(m)}) = 0 \qquad (17a)$$

For spherically symmetric potentials one can write

$$\left.\begin{aligned} Z_L^{A(B)}(E,\vec{r}) &= Y_L(\hat{r}) Z_L^{A(B)}(E,r) \\ J_L^{A(B)}(E,\vec{r}) &= Y_L(\hat{r}) = \alpha_L^{A(B)}(E,r) \end{aligned}\right\} \qquad (17b)$$

where $Z_L^{A(B)}$ and $\alpha_L^{A(B)}$ are radial wave functions and are normalized such that for $r \geq r_m$ they join smoothly to

$$Z_L^{A(B)}(\chi,\vec{r}) = J_L(\chi,r)\left[t_{A(B)}^L\right]^{-1} - i.\chi.h_L(\chi,r) \qquad (17c)$$

$$\alpha^{A(B)}(\chi,r) = J_L(\chi,r)$$

The regular wave function $Z_L^{A(B)}$ can be written as

$$Z_L^{A(B)} = \phi_L^{A(B)}(E)\psi_L^{A(B)}(r,E) \quad (17d)$$

where $\phi_L^{A(B)}(E)$ is an energy dependent renormalization factor independent of r such that $\psi_L^{A(B)} \to r^L$ for $r \to 0$.

Equations (16) and (14) are exact within the muffin-tin approximation and can be used to obtain the Green's function for a cluster of atoms[21]. However, here we are interested in an infinite system and would like to obtain ensemble average of $G(E,\vec{r},\vec{r}')$. The ensemble averages of equation (16) are related to the ensemble averages of the path operators, which will be determined by invoking the CPA. The CPA condition that the average scattering from each site must be zero, can be expressed in terms of Green's function operator as

$$x.<G>_A + y.<G>_B = G \quad (18)$$

where $<G>_{A(B)}$ denotes the restricted site average of G when zeroth site is occupied by an A (B) atom. Equivalently this condition can be written in terms of path operators as

$$x.<T_{nn}>_{n=A} + y.<T_{nn}>_{n=B} = T_{nn}^C \quad (19)$$

Where $<T_{nn}>_{n=A}$ and $<T_{nn}>_{n=B}$ are the restricted site averages of T_{nn} where n^{th} site is occupied by an A (B) atom and T_{nn}^C is the nn path operator for the CPA medium. After solving equation (19), we get the KKR-CPA condition as

$$x.[D_A]_{LL'} + y.[D_B]_{LL'} = I_{LL'} \quad (20)$$

Where

$$\left[D_{A(B)}\right]_{LL'} = \left[I + T_{00}^C(t_{A(B)}^{-1} - t_C^{-1})\right]_{LL'}^{-1} \quad (21)$$

and C is used to label quantities for the CPA medium. The path operator matrix is giver by

$$\left[T_{00}^{C}\right]_{LL'} = \frac{\Omega}{(2\pi)^3}\int\left[\left[t_C^{-1}(\chi) - B_{\vec{K}}(\chi)\right]^{-1}\right]_{LL'} d\vec{K} \qquad (22)$$

where Ω is the unit cell volume. We note that T_{00}^{C} involves a complicated Brillouin zone integration. Though $B_{\vec{K}}(\chi)$ is an off-diagonal matrix, for cubic symmetry the integral reduces to a diagonal matrix[3] for $1 \le 2$. By simple manipulations, one can reduce the KKR-CPA equation (20) with the help of equation (21) to a computationally simpler form as

$$\left[t_C^{-1}\right]_{LL'} = \left[x.t_A^{-1} + y.t_B^{-1} + (t_C^{-1} - t_A^{-1})T_{00}^{C}(t_C^{-1} - t_B^{-1})\right]_{LL'} \qquad (23)$$

This equation can be solved only by a numerical iterative method. The input for iterating this equation is t_A^{-1}, t_B^{-1} via equation (10) and $B_{\vec{K}}(\chi)$ via equation (15).

Charge Self-Consistent KKR-CPA

For full charge self-consistency, the charge densities in A and B cells are needed. These can be obtained, once the restricted site averages of the Green's function $< G(E,\vec{r},\vec{r}\,') >_{A(B)}$ are known as

$$\rho_{A(B)}(\vec{r}) = -(1/\pi)\int_{-\infty}^{E_F} \mathrm{Im} < G(E,\vec{r},\vec{r}\,') >_{A(B)} dE \qquad (24)$$

where $\rho_{A(B)}$ denotes the charge density associated with an A (B) cell. Equation (24) can be further rewritten as

$$\rho_{A(B)}(\vec{r}) = -(1/\pi)\int_{-\infty}^{E_F} \mathrm{Im}\ tr\left[Z^{A(B)}(\vec{r})Z^{A(B)}(\vec{r}\,')D_{A(B)}T_{00}^{C}\right]_{LL'} dE \qquad (25)$$

For the charge self-consistent KKR-CPA, the KKR-CPA equation (23) is first solved for given potentials of A and B atoms.

The new charge densities ρ_A and ρ_B are calculated from equation (25). New potentials are then calculated by using the local density approximation of the density-functional theory as has been described. The new potentials are used to solve the KKR-CPA equation and again a new set of potentials is calculated. We iterate this process until the potentials and charge densities get converged.

To calculate the density of state and component density of states, we define the matrices F^A and F^B as

$$F_{LL'}^{A(B)} = \int_{\Omega} Z_L^{A(B)}(\vec{r}) Z_{L'}^{A(B)}(\vec{r}) d\vec{r} \tag{26}$$

where Ω denotes the integral over the unit cell. Then the component density of states for the A and B atoms can be expressed as

$$\rho_{A(B)}(E) = -(1/\pi)\,\mathrm{Im} \int_{\Omega} < G(E,\vec{r},\vec{r}\,') >_{A(B)} d\vec{r}$$

or

$$\rho_{A(B)}(E) = -(1/\pi)\,\mathrm{Im}\; tr\left[F^{A(B)} D_{A(B)} T_{00}^{C}\right]_{LL'} \tag{27}$$

where *tr* denotes the trace in L-space. Since equation (27) is diagonal in L-space, it allows L decomposed symmetry components (s,p,t_{2g} and e_g) of $\rho_{A(B)}(E)$ in the alloy. The cubic symmetry reduces the number of distinct elements to four, i.e., s,p,t_{2g} and e_g. The average density of states for the alloy is calculated as

$$\rho(E) = -(1/\pi)\,\mathrm{Im}\; tr\left[\left\{x.F^A D_A + y.F^B D_B\right\} T_{00}^{C}\right]_{LL'} \tag{28}$$

References

1. **H. Ehrenreich and L.M. Schwartz,** *Solid State Physics 31*, (1976).
2. J.S. Faulkner in *Progress in Materials Science*, ed. By T. Massalski (Pergamon, New York, 1982) vol. *27*; J.S. Faulkner and G.M. Stocks, Phys. Rev. *B21*, 3222 (1980).

3. **A. Bansil,** *Electronic Band Structure and its Applications*, ed. M. Yussouff, (Springer-Verlag, Berlin, 1987), p 273; R.Prasad, Method of Electronic Structure Calculations, ed. O.R. Anderson, V. Kumar and A. Mookerjee (Singapore: World-Scientific) p 211; Indian J. pure Appl. Phys. 29, 255 (1991).
4. **L. Nordhem,** *Ann. Phys.* (Leipzig) 9, 607 and 641 (1931); F. Bassani and D. Brust, Phys. Rev. 131, 1524 (1963); H. Amar, K.H. Johnson, and C.B. Sommer's, Phys. Rev., 153 655 (1967); M.M. Pant and S. K. Joshi, Phys. Rev., 184, 635 (1969).
5. **N.F. Mott and H. Jones,** *The Theory of the Properties of Metals and Alloys* (Clarendon, Oxford, 1936); J. Friedel, Nuovo cimento suppl. 7, 287 (1958).
6. **T.L. Loucks,** *Augmented Plane Wave Method,* (Benjamin, New York, 1967).
7. **V.L. Moruzzi, J.F. Janak and A.R. Williams,** *Calculated Electronic Properties of Metals* (Pergamon, New York, 1978).
8. **A. Bansil, L.M. Schwartz and H. Ehrenreich,** *Phys. Rev.* B12, 2893 (1975).
9. **A. Bansil, H. Ehrenreich, L.M. Schwartz and R.E. Watson,** *Phys. Rev.* B9, 445 (1974).
10. **H. Asonen, M. Lindroos, M. Pessa, R. Prasad, R.S. Rao and A. Bansil,** *Phys. Rev.* B25, 7075 (1982).
11. **R. Prasad and A. Bansil,** *Phys. Rev. Letters*, 48, 113 (1982).
12. **R.S. Rao, A. Bansil, H. Asinen and M. Pessa,** *Phys. Rev.* B29, 1713 (1984).
13. **R. Prasad, S. C. Papadopoulos and A. Bansil,** *Phys. Rev.* B23 2607 (1981).
14. **R. Prasad, R. S. Rao and A. Bansil,** in *Excitations in Disordered System*, ed. M. P. Thorpe (Plenum, New York, 1982).
15. **P. Hohenberg and W. Kohn,** *Phys. Rev.* 136, B 864 (1964). S. Lundqvist and N.H. March, Theory of the Inhomogeneous Electron Gas, (Plenum, New York, 1983). ; J. Callaway and N. H. March, Solid State Physics, 38, 135 (1984).
16. **U. von Barth and L. Hedin,** *J. Phys. C: Solid state Phys.* 5, 1629 (1972).

17. **O. Gunnarsson and B. L. Lundqvist,** *Phys. Rev.* B13, 4274 (1976).

18. **S. H. Vosko and L. Wilk,** *Phys. Rev.* B 22, 3812 (1980); S. H. Vosko, L. Wilk and M. Nusair, Can. J. Phys. 58, 1200 (1980).

19. $\left[B_q(\chi)\right]_{LL'} = A_{LL'} + i.\chi.\delta_{LL'}$, Where $A_{LL'}$ is the well-known structure function of band theory.

20. **S. Kaprzyk and A. Bansil,** *Phys. Rev.* B42, 7358, (1990). A. Bansil and S. Kaprzyk, Phys. Rev. B43, 10335, (1991).

21. **H. Winter and G.M. Stocks,** *Phys. Rev.* B27, 882 (1983). G. Ries and H. Winter, J. Phys. F: Met. Phys. 9, 1589 (1979).

14

SUPERCONDUCTIVITY

S. K. Verma, S. K. Shrivastava,
Man Singh and Pushpraj Singh

Superconductivity is a phenomenon of exactly zero electrical resistance occurring in certain materials below a characteristic temperature. It was discovered by Heike Kamerlingh Onnes on April 8, 1911 in Leiden. Like ferromagnetism and atomic spectral lines, superconductivity is a quantum mechanical phenomenon. It is characterized by the Meissner effect, the complete ejection of magnetic field lines from the interior of the superconductor as it transitions into the superconducting state. The occurrence of the Meissner effect indicates that superconductivity cannot be understood simply as the idealization of *perfect conductivity* in classical physics.

The electrical resistivity of a metallic conductor decreases gradually as temperature is lowered. In ordinary conductors, such as copper or silver, this decrease is limited by impurities and other defects. Even near absolute zero, a real sample of a normal conductor shows some resistance. In a superconductor, the resistance drops abruptly to zero when the material is cooled below its critical temperature. An electric current flowing in a loop of superconducting wire can persist indefinitely with no power source.

In 1986, it was discovered that some cuprate-perovskite ceramic materials have a critical temperature above 90K (-183°C). Such a high transition temperature is theoretically impossible for a conventional superconductor, leading the materials to be termed high-temperature superconductors. Liquid nitrogen boils at 77 K, facilitating many experiments and applications that are less practical at lower temperatures. In conventional superconductors, electrons are held together in pairs by an attraction mediated by lattice phonons. The best available model of high-temperature superconductivity is still somewhat crude. There is a hypothesis that electron pairing in high-temperature superconductors is mediated by short-range spin waves known as paramagnons.

Classification

There is not just one criterion to classify superconductors. The most common are

- ***By their physical properties:*** They can be *Type I* (if their phase transition is of first order) or *Type II* (if their phase transition is of second order).
- ***By the theory to explain them:*** They can be *conventional* (if they are explained by the BCS theory or its derivatives) or *unconventional* (if not).
- ***By their critical temperature:*** They can be *high temperature* (generally considered if they reach the superconducting state just cooling them with liquid nitrogen, that is, if $T_c > 77$ K), or *low temperature* (generally if they need other techniques to be cooled under their critical temperature).
- ***By material:*** They can be chemical elements (as mercury or lead), alloys (as niobium-titanium or germanium-niobium or Niobium nitride), ceramics (as YBCO or the magnesium diboride), or organic superconductors (as fullerenes or carbon nanotubes, though these examples technically might be included among the chemical elements as they are composed entirely of carbon).

Elementary Properties of Superconductors

Most of the physical properties of superconductors vary from material to material, such as the heat capacity and the critical temperature, critical field, and critical current density at which superconductivity is destroyed.

On the other hand, there is a class of properties that are independent of the underlying material. For instance, all superconductors have *exactly* zero resistivity to low applied currents when there is no magnetic field present or if the applied field does not exceed a critical value. The existence of these "universal" properties implies that superconductivity is a thermodynamic phase, and thus possesses certain distinguishing properties which are largely independent of microscopic details.

Zero Electrical DC Resistance

Electric cables for accelerators at CERN: top, regular cables for LEP; bottom, superconducting cables for the LHC

The simplest method to measure the electrical resistance of a sample of some material is to place it in an electrical circuit in series with a current source I and measure the resulting voltage V across the sample. The resistance of the sample is given by Ohm's law as $R = V/I$. If the voltage is zero, this means that the resistance is zero.

Superconductors are also able to maintain a current with no applied voltage whatsoever, a property exploited in superconducting electromagnets such as those found in MRI machines. Experiments have demonstrated that currents in superconducting coils can persist for years without any measurable degradation. Experimental evidence points to a current lifetime of at least 100,000 years. Theoretical estimates for the lifetime of a persistent current can exceed the estimated lifetime of the universe, depending on the wire geometry and the temperature.

In a normal conductor, an electric current may be visualized as a fluid of electrons moving across a heavy ionic lattice. The electrons are constantly colliding with the ions in the lattice, and during each collision some of the energy carried by the current is absorbed by the lattice and converted into heat, which is essentially the vibrational kinetic energy of the lattice ions. As a result, the energy carried by the current is constantly being dissipated. This is the phenomenon of electrical resistance.

The situation is different in a superconductor. In a conventional superconductor, the electronic fluid cannot be resolved into individual electrons. Instead, it consists of bound *pairs* of electrons known as Cooper pairs. This pairing is caused by an attractive force between electrons from the exchange of phonons. Due to quantum mechanics, the energy spectrum of this Cooper pair fluid possesses an *energy gap*, meaning there is a minimum amount of energy ΔE that must be supplied in order to excite the fluid. Therefore, if ΔE is larger than the thermal energy of the lattice, given by kT, where k is Boltzmann's constant and T is the temperature, the fluid will not be scattered by the lattice. The Cooper pair fluid is thus a superfluid, meaning it can flow without energy dissipation.

In a class of superconductors known as type II superconductors, including all known high-temperature superconductors, an extremely small amount of resistivity appears at temperatures not too far below the nominal superconducting transition when an electric current is applied in conjunction with a strong magnetic field, which may be caused by the electric current. This is due to the motion of vortices in the electronic superfluid, which dissipates some of the energy carried by the current. If the current is sufficiently small, the vortices are stationary, and the resistivity vanishes. The resistance due to this effect is tiny compared with that of non-superconducting materials, but must be taken into account in sensitive experiments. However, as the temperature decreases far enough below the nominal superconducting transition, these vortices can become frozen into a disordered but stationary phase known as a "vortex glass". Below this vortex glass transition temperature, the resistance of the material becomes truly zero.

Superconducting Phase Transition

Behavior of heat capacity (c_v, blue) and resistivity (ñ, green) at the superconducting phase transition

In superconducting materials, the characteristics of superconductivity appear when the temperature T is lowered below a **critical temperature** T_c. The value of this critical temperature varies from material to material. Conventional superconductors usually have critical temperatures ranging from around 20 K to less than 1 K. Solid mercury, for example, has a critical temperature of 4.2 K. As of 2009, the highest critical temperature found for a conventional superconductor is 39 K for magnesium diboride (MgB_2), although this material displays enough exotic properties that there is some doubt about classifying it as a "conventional" superconductor. Cuprate superconductors can have much higher critical temperatures: $YBa_2Cu_3O_7$, one of the first cuprate superconductors to be discovered, has a critical temperature of 92 K, and mercury-based cuprates have been found with critical temperatures in excess of 130 K. The explanation for these high critical temperatures remains unknown. Electron pairing due to phonon exchanges explains superconductivity in conventional superconductors, but it does not explain superconductivity in the newer superconductors that have a very high critical temperature.

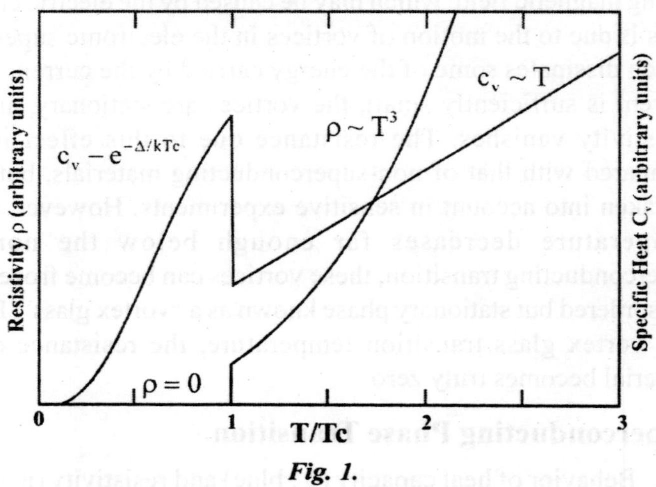

Fig. 1.

Similarly, at a fixed temperature below the critical temperature, superconducting materials cease to superconduct when an external magnetic field is applied which is greater than the *critical magnetic field*. This is because the Gibbs free energy of the superconducting phase increases quadratically with the magnetic field while the free energy of the normal phase is roughly independent of the magnetic field. If the material superconducts in the absence of a field, then the superconducting phase free energy is lower than that of the normal phase and so for some finite value of the magnetic field (proportional to the square root of the difference of the free energies at zero magnetic field) the two free energies will be equal and a phase transition to the normal phase will occur. More generally, a higher temperature and a stronger magnetic field lead to a smaller fraction of the electrons in the superconducting band and consequently a longer London penetration depth of external magnetic fields and currents. The penetration depth becomes infinite at the phase transition.

The onset of superconductivity is accompanied by abrupt changes in various physical properties, which is the hallmark of a phase transition. For example, the electronic heat capacity is proportional to the temperature in the normal (non-superconducting) regime. At the superconducting transition, it suffers a discontinuous jump and thereafter ceases to be linear. At low temperatures, it varies instead as $e^{-\alpha/T}$ for some constant, α. This exponential behavior is one of the pieces of evidence for the existence of the energy gap.

The order of the superconducting phase transition was long a matter of debate. Experiments indicate that the transition is second-order, meaning there is no latent heat. However in the presence of an external magnetic field there is latent heat, as a result of the fact that the superconducting phase has a lower entropy below the critical temperature than the normal phase. It has been experimentally demonstrated that, as a consequence, when the magnetic field is increased beyond the critical field, the resulting phase transition leads to a decrease in the temperature of the superconducting material.

Calculations in the 1970s suggested that it may actually be weakly first-order due to the effect of long-range fluctuations in the electromagnetic field. In the 1980s it was shown theoretically with the help of a disorder field theory, in which the vortex lines of the superconductor play a major role, that the transition is of second order within the type II regime and of first order (i.e., latent heat) within the type I regime, and that the two regions are separated by a tricritical point. The results were confirmed by Monte Carlo computer simulations.

Meissner Effect

When a superconductor is placed in a weak external magnetic field H, and cooled below its transition temperature, the magnetic field is ejected. The Meissner effect does not cause the field to be completely ejected but instead the field penetrates the superconductor but only to a very small distance, characterized by a parameter λ, called the London penetration depth, decaying exponentially to zero within the bulk of the material. The Meissner effect is a defining characteristic of superconductivity. For most superconductors, the London penetration depth is on the order of 100 nm.

The Meissner effect is sometimes confused with the kind of diamagnetism one would expect in a perfect electrical conductor: according to Lenz's law, when a *changing* magnetic field is applied to a conductor, it will induce an electric current in the conductor that creates an opposing magnetic field. In a perfect conductor, an arbitrarily large current can be induced, and the resulting magnetic field exactly cancels the applied field.

The Meissner effect is distinct from this—it is the spontaneous expulsion which occurs during transition to superconductivity. Suppose we have a material in its normal state, containing a constant internal magnetic field. When the material is cooled below the critical temperature, we would observe the abrupt expulsion of the internal magnetic field, which we would not expect based on Lenz's law.

The Meissner effect was given a phenomenological explanation by the brothers Fritz and Heinz London, who showed that the electromagnetic free energy in a superconductor is minimized provided

$$\nabla^2 H = 1^{-2} H$$

where H is the magnetic field and λ is the London penetration depth.

This equation, which is known as the London equation, predicts that the magnetic field in a superconductor decays exponentially from whatever value it possesses at the surface.

A superconductor with little or no magnetic field within it is said to be in the Meissner state. The Meissner state breaks down when the applied magnetic field is too large. Superconductors can be divided into two classes according to how this breakdown occurs. In Type I superconductors, superconductivity is abruptly destroyed when the strength of the applied field rises above a critical value H_c. Depending on the geometry of the sample, one may obtain an intermediate state consisting of a baroque pattern of regions of normal material carrying a magnetic field mixed with regions of superconducting material containing no field. In Type II superconductors, raising the applied field past a critical value H_{c1} leads to a mixed state (also known as the vortex state) in which an increasing amount of magnetic flux penetrates the material, but there remains no resistance to the flow of electric current as long as the current is not too large. At a second critical field strength H_{c2}, superconductivity is destroyed. The mixed state is actually caused by vortices in the electronic superfluid, sometimes called fluxons because the flux carried by these vortices is quantized. Most pure elemental superconductors, except niobium, technetium, vanadium and carbon nanotubes, are Type I, while almost all impure and compound superconductors are Type II.

London Moment

Conversely, a spinning superconductor generates a magnetic field, precisely aligned with the spin axis. The effect, the London moment, was put to good use in Gravity Probe B. This experiment measured the magnetic fields of four superconducting gyroscopes to determine their spin axes. This was critical to the experiment since it is one of the few ways to accurately determine the spin axis of an otherwise featureless sphere.

Theories of Superconductivity

Since the discovery of superconductivity, great efforts have been devoted to finding out how and why it works. During the 1950s, theoretical condensed matter physicists arrived at a solid understanding of "conventional" superconductivity, through a pair of remarkable and important theories: the phenomenological Ginzburg-Landau theory (1950) and the microscopic BCS theory (1957). Generalizations of these theories form the basis for understanding the closely related phenomenon of superfluidity, because they fall into the Lambda transition universality class, but the extent to which similar generalizations can be applied to unconventional superconductors as well is still controversial. The four-dimensional extension of the Ginzburg-Landau theory, the Coleman-Weinberg model, is important in quantum field theory and cosmology.

History of Superconductivity

Superconductivity was discovered on April 8, 1911 by Heike Kamerlingh Onnes, who was studying the resistance of solid mercury at cryogenic temperatures using the recently-produced liquid helium as a refrigerant. At the temperature of 4.2 K, he observed that the resistance abruptly disappeared. In the same experiment, he also observed the superfluid transition of helium at 2.2 K, without recognizing its significance. (The precise date and circumstances of the discovery were only reconstructed a century later, when Onnes's notebook was found.) In subsequent decades, superconductivity was observed in several other materials. In 1913, lead was found to superconduct at 7 K, and in 1941 niobium nitride was found to superconduct at 16 K.

The next important step in understanding superconductivity occurred in 1933, when Meissner and Ochsenfeld discovered that superconductors expelled applied magnetic fields, a phenomenon which has come to be known as the Meissner effect. In 1935, F. and H. London showed that the Meissner effect was a consequence of the minimization of the electromagnetic free energy carried by superconducting current.

In 1950, the phenomenological Ginzburg-Landau theory of superconductivity was devised by Landau and Ginzburg. This theory, which combined Landau's theory of second-order phase transitions with a Schrödinger-like wave equation, had great success in explaining the macroscopic properties of superconductors. In particular, Abrikosov showed that Ginzburg-Landau theory predicts the division of superconductors into the two categories now referred to as Type I and Type II. Abrikosov and Ginzburg were awarded the 2003 Nobel Prize for their work (Landau had received the 1962 Nobel Prize for other work, and died in 1968).

Also in 1950, Maxwell and Reynolds *et al.* found that the critical temperature of a superconductor depends on the isotopic mass of the constituent element. This important discovery pointed to the electron-phonon interaction as the microscopic mechanism responsible for superconductivity.

The complete microscopic theory of superconductivity was finally proposed in 1957 by Bardeen, Cooper and Schrieffer. Independently, the superconductivity phenomenon was explained by Nikolay Bogolyubov. This BCS theory explained the superconducting current as a superfluid of Cooper pairs, pairs of electrons interacting through the exchange of phonons. For this work, the authors were awarded the Nobel Prize in 1972.

The BCS theory was set on a firmer footing in 1958, when Bogolyubov showed that the BCS wavefunction, which had originally been derived from a variational argument, could be obtained using a canonical transformation of the electronic Hamiltonian. In 1959, Lev Gor'kov showed that the BCS theory reduced to the Ginzburg-Landau theory close to the critical temperature.

In 1962, the first commercial superconducting wire, a niobium-titanium alloy, was developed by researchers at Westinghouse, allowing the construction of the first practical superconducting magnets. In the same year, Josephson made the important theoretical prediction that a supercurrent can flow between two pieces of superconductor separated by a thin layer of insulator. This phenomenon, now called the Josephson effect, is exploited by superconducting devices such as SQUIDs. It is used in the most accurate available measurements of the magnetic flux quantum $\Phi_0 = \frac{h}{2e}$, and thus (coupled with the quantum Hall resistivity) for Planck's constant *h*. Josephson was awarded the Nobel Prize for this work in 1973.

In 2008, it was discovered that the same mechanism that produces superconductivity could produce a superinsulator state in some materials, with almost infinite electrical resistance.

High-Temperature Superconductivity

Until 1986, physicists had believed that BCS theory forbade superconductivity at temperatures above about 30 K. In that year, Bednorz and Müller discovered superconductivity in a lanthanum-based cuprate perovskite material, which had a transition temperature of 35 K (Nobel Prize in Physics, 1987). It was soon found that replacing the lanthanum with yttrium (i.e., making YBCO) raised the critical temperature to 92 K, which was important because liquid nitrogen could then be used as a refrigerant (the boiling point of nitrogen is 77 K at atmospheric pressure). This is important commercially because liquid nitrogen can be produced cheaply on-site from air, and is not prone to some of the problems (for instance solid air plugs) of helium in piping. Many other cuprate superconductors have since been discovered, and the theory of superconductivity in these materials is one of the major outstanding challenges of theoretical condensed matter physics.

From about 1993, the highest temperature superconductor was a ceramic material consisting of thallium, mercury, copper, barium, calcium and oxygen ($HgBa_2Ca_2Cu_3O_{8+ä}$) with T_c = 138 K.

In February 2008, an iron-based family of high-temperature superconductors was discovered. Hideo Hosono, of the Tokyo Institute of Technology, and colleagues found lanthanum oxygen fluorine iron arsenide ($LaO_{1-x}F_xFeAs$), an oxypnictide that superconducts below 26 K. Replacing the lanthanum in $LaO_{1-x}F_xFeAs$ with samarium leads to superconductors that work at 55 K.

Crystal Structure of High-Temperature Ceramic Superconductors

The structure of a high-T_c superconductor is closely related to perovskite structure, and the structure of these compounds has been described as a distorted, oxygen deficient multi-layered perovskite structure. One of the properties of the crystal structure of oxide superconductors is an alternating multi-layer of CuO_2 planes with superconductivity taking place between these layers. The more layers of CuO_2 the higher T_c. This structure causes a large anisotropy in normal conducting and superconducting properties, since electrical currents are carried by holes induced in the oxygen sites of the CuO_2 sheets. The electrical conduction is highly anisotropic, with a much higher conductivity parallel to the CuO_2 plane than in the perpendicular direction. Generally, Critical temperatures depend on the chemical compositions, cations substitutions and oxygen content. They can be classified as superstripes; i.e., particular realizations of superlattices at atomic limit made of superconducting atomic layers, wires, dots separated by spacer layers, that gives multiband and multigap superconductivity.

YBaCuO Superconductors

The first superconductor found with $T_c > 77$ K (liquid nitrogen boiling point) is yttrium barium copper oxide ($YBa_2Cu_3O_{7-x}$), the proportions of the 3 different metals in the $YBa_2Cu_3O_7$ superconductor are in the mole ratio of 1 to 2 to 3 for yttrium to barium to copper respectively. Thus, this particular superconductor is often referred to as the 123 superconductor.

The unit cell of $YBa_2Cu_3O_7$ consists of three pseudocubic elementary perovskite unit cells. Each perovskite unit cell contains a Y or Ba atom at the center: Ba in the bottom unit cell, Y in the middle one, and Ba in the top unit cell. Thus, Y and Ba are stacked in the sequence [Ba–Y–Ba] along the c-axis. All corner sites of the unit cell are occupied by Cu, which has two different coordinations, Cu(1) and Cu(2), with respect to oxygen. There are four possible crystallographic sites for oxygen: O(1), O(2), O(3) and O(4). The coordination polyhedra of Y and Ba with respect to oxygen are different. The tripling of the perovskite unit cell leads to nine oxygen atoms, whereas $YBa_2Cu_3O_7$ has seven oxygen atoms and, therefore, is referred to as an oxygen-deficient perovskite structure. The structure has a stacking of different layers: (CuO)(BaO)(CuO_2)(Y)(CuO_2)(BaO)(CuO). One of the key feature of the unit cell of $YBa_2Cu_3O_{7-x}$ (YBCO) is the presence of two layers of CuO_2. The role of the Y plane is to serve as a spacer between two CuO_2 planes. In YBCO, the Cu–O chains are known to play an important role for superconductivity. T_c is maximal near 92 K when $x \approx 0.15$ and the structure is orthorhombic. Superconductivity disappears at $x \approx 0.6$, where the structural transformation of YBCO occurs from orthorhombic to tetragonal.

Bi-, Tl- and Hg-based High-T_c *Superconductors*

The crystal structure of Bi-, Tl- and Hg-based high-T_c superconductors are very similar. Like YBCO, the perovskite-type feature and the presence of CuO_2 layers also exist in these superconductors. However, unlike YBCO, Cu–O chains are not present in these superconductors. The YBCO superconductor has an orthorhombic structure, whereas the other high-T_c superconductors have a tetragonal structure.

The **Bi–Sr–Ca–Cu–O** system has three superconducting phases forming a homologous series as $Bi_2Sr_2Ca_{n-1}Cu_nO_{4+2n+x}$ (n = 1, 2 and 3). These three phases are Bi-2201, Bi-2212 and Bi-2223, having transition temperatures of 20, 85 and 110 K, respectively, where the numbering system represent number of atoms for Bi, Sr, Ca and Cu respectively. The two phases have a tetragonal structure which consists of two sheared crystallographic unit cells. The unit cell of these phases has double Bi–O planes which are stacked in a way that the Bi atom of one plane sits below the oxygen atom of the next consecutive plane. The Ca atom forms a layer within the interior of the CuO_2 layers in both Bi-2212 and Bi-2223; there is no Ca layer in the Bi-2201 phase. The three phases differ with each other in the number of CuO_2 planes; Bi-2201, Bi-2212 and Bi-2223 phases have one, two and three CuO_2 planes, respectively. The c axis of these phases increases with the number of CuO_2 planes. The coordination of the Cu atom is different in the three phases. The Cu atom forms an octahedral coordination with respect to oxygen atoms in the 2201 phase, whereas in 2212, the Cu atom is surrounded by five oxygen atoms in a pyramidal arrangement. In the 2223 structure, Cu has two coordinations with respect to oxygen: one Cu atom is bonded with four oxygen atoms in square planar configuration and another Cu atom is coordinated with five oxygen atoms in a pyramidal arrangement.

Tl–Ba–Ca–Cu–O superconductor: The first series of the Tl-based superconductor containing one Tl–O layer has the general formula $TlBa_2Ca_{n-1}Cu_nO_{2n+3}$, whereas the second series containing two Tl–O layers has a formula of $Tl_2Ba_2Ca_{n-1}Cu_nO_{2n+4}$ with $n = 1, 2$ and 3. In the structure of $Tl_2Ba_2CuO_6$ (Tl-2201), there is one CuO_2 layer with the stacking sequence (Tl–O) (Tl–O) (Ba–O) (Cu–O) (Ba–O) (Tl–O) (Tl–O). In $Tl_2Ba_2CaCu_2O_8$ (Tl-2212), there are two Cu–O layers with a Ca layer in between. Similar to the $Tl_2Ba_2CuO_6$ structure, Tl–O layers are present outside the Ba–O layers. In $Tl_2Ba_2Ca_2Cu_3O_{10}$ (Tl-2223), there are three CuO2 layers enclosing Ca layers between each of these. In Tl-based superconductors, T_c is found to increase with the increase in CuO_2 layers. However, the value of T_c decreases after four CuO_2 layers in $TlBa_2Ca_{n-1}Cu_nO_{2n+3}$, and in the $Tl_2Ba_2Ca_{n-1}Cu_nO_{2n+4}$ compound, it decreases after three CuO_2 layers.

Hg–Ba–Ca–Cu–O superconductor: The crystal structure of $HgBa_2CuO_4$ (Hg-1201), $HgBa_2CaCu_2O_6$ (Hg-1212) and $HgBa_2Ca_2Cu_3O_8$ (Hg-1223) is similar to that of Tl-1201, Tl-1212 and Tl-1223, with Hg in place of Tl. It is noteworthy that the T_c of the Hg compound (Hg-1201) containing one CuO_2 layer is much larger as compared to the one-CuO_2-layer compound of thallium (Tl-1201). In the Hg-based superconductor, T_c is also found to increase as the CuO_2 layer increases. For Hg-1201, Hg-1212 and

Hg-1223, the values of T_c are 94, 128 and 134 K respectively, as shown in table below. The observation that the T_c of Hg-1223 increases to 153 K under high pressure indicates that the T_c of this compound is very sensitive to the structure of the compound.

The simplest method for preparing high-T_c superconductors is a solid-state thermochemical reaction involving mixing, calcination and sintering. The appropriate amounts of precursor powders, usually oxides and carbonates, are mixed thoroughly using a ball mill. Solution chemistry processes such as coprecipitation, freeze-drying and sol-gel methods are alternative ways for preparing a homogenous mixture. These powders are calcined in the temperature range from 800 °C to 950 °C for several hours. The powders are cooled, reground and calcined again. This process is repeated several times to get homogenous material. The powders are subsequently compacted to pellets and sintered. The sintering environment such as temperature, annealing time, atmosphere and cooling rate play a very important role in getting good high-T_c superconducting materials. The $YBa_2Cu_3O_{7-x}$ compound is prepared by calcination and sintering of a homogenous mixture of Y_2O_3, $BaCO_3$ and CuO in the appropriate atomic ratio. Calcination is done at 900–950 °C, whereas sintering is done at 950 °C in an oxygen atmosphere. The oxygen stoichiometry in this material is very crucial for obtaining a superconducting $YBa_2Cu_3O_{7-x}$ compound. At the time of sintering, the semiconducting tetragonal $YBa_2Cu_3O_6$ compound is formed, which, on slow cooling in oxygen atmosphere, turns into superconducting $YBa_2Cu_3O_{7-x}$. The uptake and loss of oxygen are reversible in $YBa_2Cu_3O_{7-x}$. A fully oxidized orthorhombic $YBa_2Cu_3O_{7-x}$ sample can be transformed into tetragonal $YBa_2Cu_3O_6$ by heating in a vacuum at temperature above 700 °C.

The preparation of Bi-, Tl- and Hg-based high-T_c superconductors is difficult compared to YBCO. Problems in these superconductors arise because of the existence of three or more phases having a similar layered structure. Thus, syntactic intergrowth and defects such as stacking faults occur during synthesis and it becomes difficult to isolate a single superconducting phase. For Bi–Sr–Ca–Cu–O, it is relatively simple to prepare the Bi-2212 (T_c ~ 85 K) phase, whereas it is very difficult to prepare a single phase of Bi-2223 (T_c ~ 110 K). The Bi-2212 phase appears only after few hours of sintering at 860–870 °C, but the larger fraction of the Bi-2223 phase is formed after a long reaction time of more than a week at 870 °C. Although the substitution of Pb in the Bi–Sr–Ca–Cu–O compound has been found to promote the growth of the high-T_c phase, a long sintering time is still required.

Possible Superconductivity of The Vacuum

Maxim Chernodub of the French National Centre for Scientific Research has postulated that the vacuum can become a superconductor in magnetic fields of 10^{16} Tesla or more, at temperatures of at least a billion, perhaps billions of degrees.

Applications

Superconducting magnets are some of the most powerful electromagnets known. They are used in MRI and NMR machines, mass spectrometers, and the beam-steering magnets used in particle accelerators. They can also be used for magnetic separation, where weakly magnetic particles are extracted from a background of less or non-magnetic particles, as in the pigment industries.

In the 1950s and 1960s, superconductors were used to build experimental digital computers using cryotron switches. More recently, superconductors have been used to make digital circuits based on rapid single flux quantum technology and RF and microwave filters for mobile phone base stations.

Superconductors are used to build Josephson junctions which are the building blocks of SQUIDs (superconducting quantum interference devices), the most sensitive magnetometers known. SQUIDs are used in scanning SQUID microscopes and magnetoencephalography. Series of Josephson devices are used to realize the SI volt. Depending on the particular mode of operation, a superconductor-insulator-superconductor Josephson junction can be used as a photon detector or as a mixer. The large resistance change at the transition from the normal- to the superconducting state is used to build thermometers in cryogenic micro-calorimeter photon detectors. The same effect is used in ultrasensitive bolometers made from superconducting materials.

Other early markets are arising where the relative efficiency, size and weight advantages of devices based on high-temperature superconductivity outweigh the additional costs involved.

Promising future applications include high-performance smart grid, electric power transmission, transformers, power storage devices, electric motors (e.g. for vehicle propulsion, as in vactrains or maglev trains), magnetic levitation devices, fault current limiters, nanoscopic materials such as buckyballs, nanotubes, composite materials and superconducting magnetic refrigeration. However, superconductivity is sensitive to moving magnetic fields so applications that use alternating current (e.g. transformers) will be more difficult to develop than those that rely upon direct current.

15

EM and Microwave Absorption of Mn-Zn and Ni-Zn Ferrite

K. Chaturvedi. Bhoopendra Singh
Ashish Autiyal and Sachin Shrivastava

Introduction

In past few decades, E.M. waves played an increasing role in man's technological achievements, and composite materials capable of absorbing microwave radiation are becoming an increasingly important resource. These composites find application in shielding of electronic components, reduction of cross-talk in cell phones, stealth technology and to control heat distribution in microwave-assisted curing or in domestic cooking. The EM absorbers are in great demand in different GHz frequency ranges, such as mobile phones (0.8-1.5GHz), LAN systems (2.45, 5.0, 22.0, 60.0 GHz) and satellite broadcast systems (11.7-12.0 GHz), the EM absorbers have widely investigated to eliminate the above problems.

The information and communication technology is making a dramatic advancement, and growing as a big market in today's society like cellular phone, cordless telephone, television, computer, electronic note pad, direct satellite to home system, high speed radio LAN (Local area network), microwave oven etc., are being used every day in modern life style. Most of these radiate EM waves in microwave frequency (1-40 GHz) region. It is well known that microwaves generated from these devices have harmful effects on the people who use these devices. These effects include increase in heart beats, weakening of immune systems, rearrangement of proteins including DNA, increasing possibility of leukemia, sterility, cataract, cancer etc. There, the research for suitable EM absorbing materials to reduce harmful effects of EM waves electronic equipments or human beings, and to reduce radar signature is important.

Now days, the microwaves in higher GHz ranges are increasingly exploited by circuitry industries dealing with wireless telecommunication systems, LAN, medical equipments, radars etc. As the more electronic equipments and dense circuit packaging within the devices are increasing, troubles caused by EM noise generated from these devices increasingly reported.

EM wave absorbing technology is an important topic for military purposes and is also a rising major issue in the business field. Especially, the development of radar absorbing material (RAM), EM wave absorbing materials in the frequency region of radar, GHZ bandwidth, had been actively researched for quite a long time. The radar (radio detection and ranging) is a system that transmits EM (Electromagnetic) waves in a given direction and then detects the same waves reflected back by an obstacle in its path .The radar technology has been improved drastically with the use of high-powered large bandwidth transmitters since the World War-II, which makes the stealth technology very important for the survival of the weapon systems, such as aircrafts, warships, and missiles in modern wars. The stealth technology is a sub-discipline of ECM (electronic countermeasures) which covers a range of techniques for weapon systems in order to make them less visible or ideally invisible to the enemy's radar. For the stealth performance, the RCS (radar cross section) of weapon systems should be minimized because the distance detected by a radar is inversely proportional to the fourth root of the RCS. RCS is a measurement of the strength of the radar signal backscattered from a target object for a given incident EM wave power to describe the extent to which an object reflects an incident EM wave. There are several methods to reduce the RCS such as shaping the target, and employing RAM (radar absorbing materials) or RAS (radar absorbing structures). Microwave-absorbing materials (MAMs) have wide commercial applications. Besides being applied in electronic devices and systems of wireless antenna and cellular phones, it is used as shielding substances against EMI (electromagnetic interference), the common problems due to EMI are malfunctioning of devices, formation of ghost or false images, inconsistent radar signals etc., and due to these reasons, operation of mobile phones and other wireless devices are strictly prohibited at some specific time and places like hospitals, bank ATMs, aero planes etc. Application of MAMs in appropriate places in electronic equipments or host platforms controls the excessive self-emitting EM waves and also ensures the undisturbed functioning the equipments in presence of external EM waves. Achieving the control of self- emission and disallowing the unwanted external EM waves to enter in equipments is often referred as electromagnetic compatibility (EMC). A ferrite tile is a well known example of EM wave absorptibity and is provided on a wall of a high-rise building for the purpose of TV ghost phenomenon prevention. Traditionally, MAMs are particles of magnetic metals or alloys, and because of high specific gravity and formulation difficulty, they are restricted in application. It is hence desirable to have MAMs that are lightweight, structurally sound, and flexible and show good microwave-absorbing ability in a wide frequency range.

Advances in The Field of Microwave Absorbers

Microwave lossy materials are in use for long time in both civil and military application. W.H. Emerson reviewed the early work on EM wave absorber. He traced the earliest research work on E.M wave absorber to the mid thirties to improve the performance an antenna at 2 GHz. It was a quarter wavelength resonant type absorber prepared using the titania (TiO_2) as a high dielectric constant material to reduce the thickness, and carbon black for dielectric loss. However, the invention of radar during World War II, served as the impetus for intense research on radar absorbing materials (RAMs). The primary aim was to develop absorbers for radar camouflage. Work on RAMs was being conducted in the Germany and United States simultaneously. German took the lead to develop first RAM during the war. The one material employed was a natural rubber sheet filled with charcoal powder to cover their submarine periscopes, thus reducing their detection by radar. Another early German material RAM was developed by J. Jaumann. It was a rigid broadband material of about 3 inch thickness comprising of alternate layers of rigid plastic and resistive sheets. The Jaumann absorber effectively provided a reflection level of better than -20 dB over the frequency range of about 2 – 15 GHz. This type of absorber designed for multi-band absorption is still used today in RAS composites.

Early US work produced a material called Halpern Anti Radiation Paint (HARP) patented by Otto Halpern at MIT radiation laboratory. This was the first sprayable material consisted of aligned flakes of aluminum, copper or ferromagnetic magnetic materials in plastic or rubber binder. This could be sprayed or trowelled onto the ship structures, provided not only minimized possibility of detection by radar, but also environmental protection to the metal substrate like paint. This was again a resonant absorber having a dielectric constant ~ 20 and thickness approximately 2.0 mm, provided a reflection level of -15 dB to – 20 dB in X-band (8.2 – 12.4 GHz). Another remarkable early work of MIT radiation laboratory was a development of Salisbury screen. The construction was very simple; this was prepared by sticking the metal foil to one side of a piece of around 20.0 mm thick plywood, and Uskon cloth as resistive sheet having the resistivity 377 Ω/γ to the other side. The dielectric constant of the wood was such that resonance condition occurred near 3 GHz with this thickness. The efforts were continued, even after the WW II was over, to realize thin, broadband absorbers for two fold applications: reduction of RCS and to improve the performance of radar and communication systems by suppression of EMI.

The ferrite based EM absorbers were first explored by Y. Naaito. He studied the absorbing properties of various sintered spinel ferrites such as MnZn, NiZn, NicuZn,, NiCuZnCo, MgCuZn,

NiZnCo and NiMgZn. He discovered that metal backed ferrite absorbers have a characteristic matching thickness (d_m) for which minimum reflection occurred at a characteristic matching frequency (f_m). Interestingly, d_m was always either 8 mm or 5 mm (for ferrite containing cobalt). f_m for different ferrite was different, and it was not same as ferrite's resonance frequency (f_r). Ratio f_m / f_r depend upon the value of the μ_r' and ranges from 3 to 27, being larger for larger values of μ_r'. he also studied the ferrite- rubber composites, and found that there exist a critical ferrite to rubber weight ratio (W_c) with the properties: (*i*) no matching frequency for $W < W_c$; (*ii*) one matching frequency when $W = W_c$; (*iii*) two matching frequencies (f_{m1} and f_{m2}) for $W > W_c$; (*iv*) second matching thickness could be determined from the constancy of the product $d_{m2}.f_{m2} = 28$ to 30 GHz. mm. The author observed that the matching frequency (f_{m1}) of sintered ferrites could not exceed 2.0 GHz, but for ferrite-rubber composites, the upper limit of f_{m1} could be extended to f_{m2} in the range of 4–12 GHz. He further established that for the same thickness, ferrite absorber was superior to dielectric absorber because the former showed the larger relative bandwidth (B/f_0); defined as the ratio of the bandwidth for the standing–wave ratio 1.2 to the mean frequency of the band.

Severin and Stoll were the first to study the complex permeabilities and resonance properties of Co-Ti substituted, Ba-hexaferrite in the frequency range of 2 – 52 GHz. They had showed the way, how a hexaferrite could be tuned to resonate at desired frequencies. This had formed the basis for deploying the hexaferrites in absorber applications. James and Amin had best employed the hexaferrites to create an effective microwave absorber, by suitable choice of material parameters and thickness. They adopted two multi-layer approaches for the absorber design. In one case, a tapered layer configuration was used to stagger the material properties and layer thickness. This gave a relatively large broadband absorption of – 10 dB over the frequency range of 2 – 20 GHz, and around -20 dB in X-band (8.2 – 12.4 GHz) region in a total thickness of about 7.5 mm. In second approach, different layers of quarter wavelength resonant absorbers were used. This gave absorption of -10 dB in the narrow frequency range of 6 – 10 GHz, but the total thickness was reasonably low, about 1.8 mm.

J. Ajadmanmanjiri have studied the Ni-Zn ferrite nanoparticles with different composition 0f $Ni_{1-x}Zn_xFe_2O_4$ (x = 0, 0.1, 0.2, 0.3, 0.4). He found that Zn content has significant influence on the electromagnetic properties such as dielectric constant (ε_r'), dielectric loss tangent and complex dielectric constant (ε_r'). These values are decreases with increasing of zinc. Huang Aiping et al., have studied the electromagnetic properties of the Mn-Zn with Fe poor composition and simulate the permeability spectra of ferrites. He observed that the permeability of Fe poor Mn-Zn ferrites in comparison to that of Fe-rich is still high enough to the Ni-Zn ferrites in high frequency region.

Very recently, T. Kagotani et al., have reported the studies on enhancement of GHz electromagnetic wave absorption characteristics in aligned M-type barium ferrite $Ba_{1-x}La_xZn_x(Me_{0.5}Mn_{0.5})_yFe_{12-x-y}O_{19}$ (x = 0.0 - 0.5 & y = 1.0 – 3.0; Me: Zr, Sn) by metal substitution. They could not notice any measurable effect of La and Zn substitution. It seems that the additive elements did not substitute for Fe^{3+} in the structure of ferrite, but were mixed as inclusions in the ferrite matrix. However, in a simple composition $Ba(Zr_{0.5}Mn_{0.5})_{1.3}Fe_{10.7}O_{19}$, they obtained rerromagnetic permeabilities, $\mu_r' = 7.0$ and $\mu_r'' = 6.5$; and very sharp reflection loss of -20 dB almost on a single frequency at 10 GHz in absorber thickness of just 0.58 mm. This has proved that if permeability values are large, matching thickness can be reduced; provided these large values existed for broad frequency range, other it renders only single or very narrow frequency loss which does not have much application except as a notch filter. M. R. Meshram et al., have studied the electromagnetic and microwave absorption properties of the M-type Ba-hexaferrites- epoxy paints with two hexaferrites compositions: $BaCo_{0.5\delta}Ti_{0.5\delta}Mn_{0.1}Fe_{11.87-\delta}O_{19}$ and $BaTi_\delta Mn_\delta Fe_{12-2\delta}O_{19}$ at $\delta = 1.6$; in the X-band frequency range of 8.2 to 12.4 GHz.

Praveen Singh et al., have studied the complex permittivity, permeability, and microwave absorption properties of Co-Ti substituted Calcium hexaferrite–epoxy composites with the ferrite composition $[(CaCo_xTi_xFe_{12-2x}O_{19})_{96.0}(La2O3)_{4.0}]$ for x= 0.0 to 1.0 in step of 0.2, in the frequency range from 8.0to 12.4 GHz. They obtained maximum permeability values, $\mu_r' = 1.3$ and $\mu_r = 0.6$; and maximum permittivity values $\varepsilon_r' = 4.2$ and $\varepsilon_r^2 = 1.5$ for the ferrite composition at x = 0.2. In another study same group has carried out similar studies on Ni-Ti substituted Calcium hexaferrite – epoxy composite. In this study, they obtained maximum permeability values, $\mu_r' = 1.2$ and $\mu_r'' = 0.6$; and maximum permittivity values $\varepsilon_r' = 4.1$ and $\varepsilon_r^2 = 1.0$ for the ferrite composition at x = 0.6. But the best absorption properties were obtained for ferrite composition at x = 0.4; that is the maximum reflection loss of -30 dB at 9.5 GHz with associated bandwidth (full width at half maximum) of only 0.5 GHz in a sample thickness of 4.15 mm. As compared to their previous study, results of this study are somewhat inferior.

S. K. Kwon et al., have studied the microwave absorption properties of carbon black –silicone rubber blend for varying carbon contents and sample thickness in the frequency range of 2–18 GHz. They obtained reflection loss of -10 dB over the frequency range of 9–13 GHz for the best sample containing 10 % carbon black by weight in a sample thickness of 2.0 mm. This sample showed permittivity values, $\varepsilon_r' = 15$ and $\varepsilon_r'' = 4.0$. The permittivity values were found to increase with increased carbon contents.

The patent by Y. Nikawa et al., describes the manufacturing of the flexible light weigth carbon based microwave absorber. They have prepared these absorbers using both amorphous carbon black and crystalline graphite powders in silicone resin. Authors have emphasized on the dispersion of carbon particles in the insulating matrix, and used a special method to ensure their proper dispersion. They have obtained $\varepsilon_r' = 10$ and $\varepsilon_r'' = 4.5$ at 10 GHz, for the sample the containing 31 % carbon by weight. This sample has showed the -20 dB reflection loss, over narrow frequency range of 9.2 to 9.8 GHz; for the sample thickness of 2.6 mm.

Origin of Electromagnetic Waves

There are various forms of energy which include heat, light, sound, electricity, and chemical, energy radiated in the form of a wave. Wave, in physics, is the transfer of energy by the regular vibration or oscillatory motion of either of some material medium or by the variation in magnitude of the field vectors of an electromagnetic field as a result of the motion of electric charges. A moving charge gives rise to a magnetic field, and if the motion is changing, then the magnetic field varies and in turn produces an electric field. These interacting electric and magnetic fields are at right angles to one another and also to the direction of propagation of the energy. Thus, an electromagnetic wave is a transverse wave. If the direction of the electric field is constant, the wave is said to be polarized. Electromagnetic radiation does not require a material medium and can travel through a vacuum. The theory of electromagnetic radiation was developed by James Clerk Maxwell and published in 1865. He showed that the speed of propagation of electromagnetic radiation should be identical with that of light, visible electromagnetic radiation. Heinrich Hertz verified Maxwell's prediction through the discovery of radio waves, also known as hertzian waves. Light is a type of electromagnetic radiation, occupying only a small portion of the possible spectrum, arrangement or display of light or other form of radiation separated according to wavelength, frequency, energy, or some other property. Beams of charged particles can be separated into a spectrum according to mass in a mass spectrometer. The various types of electromagnetic radiation differ only in wavelength and frequency; they are alike in all other respects. The possible sources of electromagnetic radiation are directly related to wavelength: long radio waves are produced by large antennas such as those used by broadcasting stations; much shorter visible light waves are produced by the motions of charges within atoms, basic unit of matter ; more properly, the smallest unit of a chemical element having the properties of that element.

Generally, EM radiation is classified by wavelength into electrical energy, radio, microwave, infrared, the visible region we perceive as light, ultraviolet, X-rays and gamma rays. The behavior of EM radiation depends on its wavelength. Higher frequencies have shorter wavelengths, and lower frequencies have longer wavelengths. When EM radiation interacts with single atoms and molecules, its behavior depends on the amount of energy per quantum it carries.

Fundamentals of FM Waves

Electromagnetic (EM) waves are created by time-varying currents and charges. Their interactions with materials obey the boundary conditions of Maxwell's equations. EM waves can be guided by structures (transmission lines) or by free space. An antenna is a material structure that directs EM fields from a source into space, or, by reciprocity, from space to a receiver. The shape and size of the antenna controls the transition from the near field to the far field. The propagation of electromagnetic wave is shown in fig. 1. Near-field behavior is most clearly seen surrounding small antennas; the electric dipole is a capacitive object. Near field behavior is shown in fig. 2.

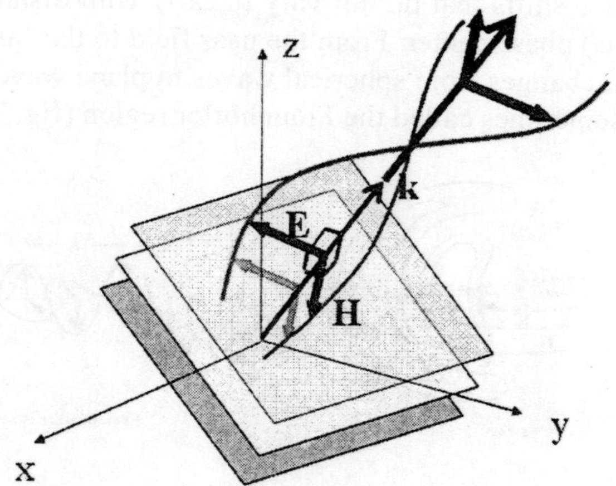

Fig. 1. Propagation of Electromagnetic Waves

The near field consists of the reactive near field, also known as the quasi-static near field, and the radiating near field, also known as the Fresnel zone or Fresnel region. In the quasi-static near field we see fields that strongly resemble the electrostatic fields of a charge dipole for a dipole antenna and the fields of a magnetic dipole for a loop antenna. In large antennas the quasi-static field can be seen near edges.

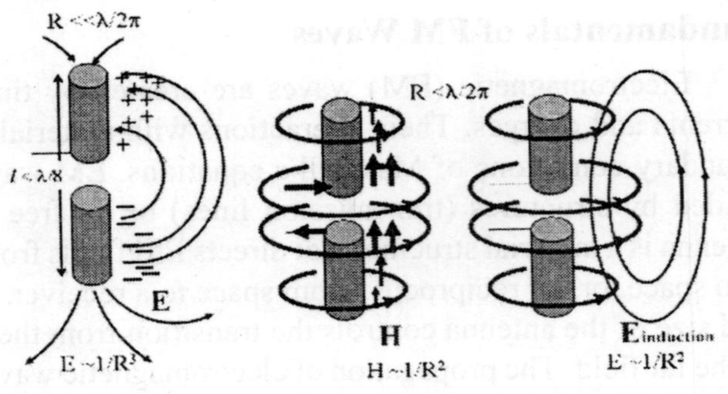

Fig. 2. Near Field Behavior of EM Waves

In the Fresnel zone the waves are clearly not plane and may have phase shifts that do not vary linearly with distance from a (fictitious) phase center. From the near field to the far field, EM radiation changes from spherical waves to plane waves. The far-field is sometimes called the Fraunhoffer region (fig. 3).

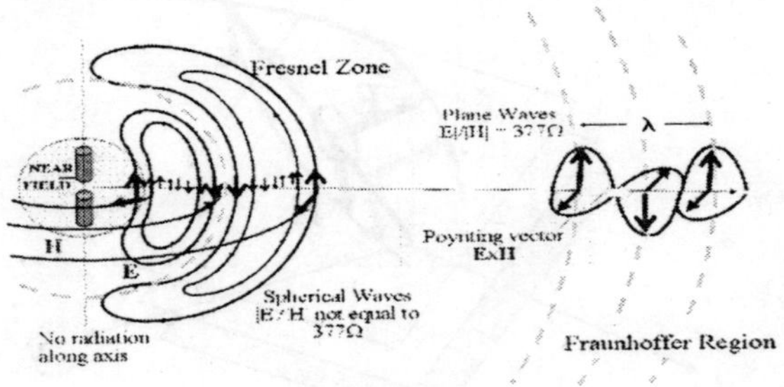

Fig. 3. Near Field to Far Field Behavior of EM Waves

The basic structure of matter involves charged particles bound together in many different ways. When electromagnetic radiation is incident on matter, it causes the charged particles to oscillate and gain energy. The ultimate fate of this energy depends on the situation. It could be immediately re-radiated and appear as scattered, reflected, or transmitted radiation. It may also get dissipated into other microscopic motions within the matter, coming to thermal equilibrium and manifesting itself as thermal energy in the material. With a few exceptions such as fluorescence, harmonic generation, photochemical reactions and the photovoltaic effect, absorbed electromagnetic radiation simply deposits its energy by heating the material. This happens both for infrared and non-infrared radiation. Intense radio waves can thermally burn living tissue and can cook food. In addition to infrared lasers, sufficiently intense visible and ultraviolet lasers can also easily set paper afire. Ionizing electromagnetic radiation can create high-speed electrons in a material and break chemical bonds, but after these electrons collide many times with other atoms in the material eventually most of the energy gets downgraded to thermal energy, this whole process happening in a tiny fraction of a second. That infrared radiation is a form of heat and other electromagnetic radiation is not, is a widespread misconception in physics. Any electromagnetic radiation can heat a material when it is absorbed.

Electromagnetic Wave Equations

Electromagnetic waves as a general phenomenon were predicted by Classical laws of electricity and magnetism, known as Maxwell's equations. If we inspect Maxwell's equations without sources (charges or currents) then we shall find that, along with the possibility of nothing happening, the theory will also admit nontrivial solutions of changing electric and magnetic fields. Beginning with Maxwell's equations for free space:

$$\nabla . E = \frac{\rho}{E_0} \tag{1}$$

$$\nabla X E = -\frac{\partial B}{\partial t} \tag{2}$$

$$\nabla . B = 0 \tag{3}$$

$$\nabla X B = \mu_0 \varepsilon_0 \frac{\partial E}{\partial t} \tag{4}$$

where, ∇ is a vector differential operator.

Maxwell's equations have unified the permittivity of free space ε_0, the permeability of free space μ_0, and the speed of light c_0. Let's consider a generic vector wave for the electric field.

$$E = E_0 f(k, x - c_0 t) \tag{5}$$

Here E_0 is the constant amplitude, f is any second differentiable function, $\hat{k}$ is a unit vector in the direction of propagation, and x is a position vector. We observe that $f(\hat{k}, x - c_0 t)$ is a generic solution to the wave equation. In other words

$$\nabla^2 f(\hat{k}, x - c_0 t) = \frac{1}{c_0^2 \partial t^2} f(\hat{k}, x - c_0 t) \tag{6}$$

for a generic wave traveling in the k direction.

Not only are the electric and magnetic field waves traveling at the speed of light, but they have a special restricted orientation and proportional magnitudes, $E_0 = c_0 B_0$, which can be seen immediately from the Poynting vector. The electric field, magnetic field, and direction of wave propagation are all orthogonal, and the wave propagates in the same direction as ExB.

From the viewpoint of an electromagnetic wave traveling forward, the electric field might be oscillating up and down, while the magnetic field oscillates right and left; but this picture can be rotated with the electric field oscillating right and left and the magnetic field oscillating down and up. This is a different solution that is traveling in the same direction. This arbitrariness in the orientation with respect to propagation direction is known as polarization.

The electromagnetic field in wave propagation in a medium can be solved from the Maxwell's equations:

$$\nabla \times E = -\frac{\partial B}{\partial t} \tag{7}$$

$$\nabla \times H = J + \frac{\partial D}{\partial t} \tag{8}$$

$$\nabla \times D = \rho \quad (9)$$

$$\nabla.B = 0 \quad (10)$$

$$\nabla.J = -\frac{\partial \rho}{\partial t} \quad (11)$$

The electrical properties of the medium (σ, ε, μ) offer consecutive relations.

$D = \varepsilon E = (\varepsilon' - j \in'')E$ = Electric displacement vector

$J = \sigma E + J_i$ = Electrical current density

$\varepsilon = \varepsilon_r \varepsilon_0$ = Electric permittivity

$\mu = \mu_r \mu_0$ = Magnetic permeability

Here J_i is the externally impressed current density producing *the electromagnetic fields.* σ, ε'' and μ'' account for joule heating loss, dielectric loss and magnetic loss in the medium respectively. The loss tangent of the medium is given by

$$\tan \delta = (\omega \varepsilon'' + \sigma)/\omega \varepsilon' \quad (12)$$

For time harmonic fields (e^{jwt}) the above equations reduce to

$$\nabla \times E = -j\omega\mu \quad (13)$$

$$\nabla \times H = (\sigma + j\omega\varepsilon)E \quad (14)$$

$$\nabla.D = \rho \quad (15)$$

$$\nabla.B = 0 (16)$$

$$\nabla.J = -j\omega\rho \quad (17)$$

Since both E and H in these equations are unknown, solution for electric or magnetic fields can be obtained from wave equations involving either the E or H fields derived from equations 13 to 17.

$$\nabla^2 E + k^2 E = j\omega\mu J - \frac{\nabla\nabla.J}{j\omega\rho} \quad (18)$$

$$\nabla^2 H + k^2 H = -\nabla \times J \tag{19}$$

where, $k = \sqrt{\{-j\omega\mu(\sigma + j\omega\varepsilon)\}} = \sqrt{[-j\omega\mu\sigma + \omega^2\mu\omega]} = \alpha + j\beta$ (20)

Here k is called the propagation constant of the medium. In a medium with finite conductivity σ, a conduction current density $J = \sigma E$ will exist to give attenuation α because of joule heating. In a non-conducting lossless medium, $\sigma = 0$ and $k = \omega(\mu\varepsilon)1/2$. At frequencies below the optical region the propagation constant in a good conductor $(\sigma << \omega\varepsilon)$ becomes

$$k = (1 + j)\sqrt{(\pi f\mu\sigma)} = \alpha + j\beta \tag{21}$$

And in free space where $(\sigma \sim 0), k = \omega\left(\mu_0\varepsilon_0\right)^{\frac{1}{2}}$

Microwave can penetrate a conductor falling to 1/e of its surface value at a depth called 'skin depth' which is given by

$$\delta = \frac{1}{\sqrt{\pi f\mu\sigma}} = \frac{1}{-}\alpha \tag{22}$$

The propagating wave characterized by wave impedance in free space is

$$Z_0 = \sqrt{\mu_0/_{\varepsilon_0}} = 120\pi \text{ ohm} \tag{23}$$

and that in lossy conducting media is

$$z = (1 + j)\sqrt{\omega\mu - 2\sigma} = (1 + j)R_s \tag{24}$$

where R_S is the conductor surface resistance given by

$$R_s = \sqrt{\omega\mu - 2\sigma} \tag{25}$$

When a potential source supplies time varying currents in a conductor it emits electromagnetic energy. The energy supplied by the source is partly stored in the electric and magnetic fields inside a closed volume in the immediate vicinity of the conductor and partly propagated away as an EM wave in lossless transmission medium, such as free space, coaxial line, waveguides, etc.

For the steady state time harmonic excitation (e^{jwt}), the time-average energies stored in the electric magnetic fields in a volume are given by

$$W_e = \mathrm{Re}\frac{1}{4}\int E.D dv = \frac{\sigma\varepsilon}{4\int E^2 dV} \quad (26)$$

$$W_m = \mathrm{Re}\frac{1}{4}\int H.d V = \frac{\mu'}{4\int H^2 dV} \quad (27)$$

where peak values are used for the complex fields E and H.

For EM wave propagation, the time-average power transmitted through a unit area is called the pointing vector, P.

$$P = \frac{1}{2}(E \times H'') \quad (28)$$

The time average complex power flow across a surface S is, therefore

$$S = \frac{1}{2}\int (E \times H'').dS \quad (29)$$

Classification of Materials

Materials can be classified according to their permittivity and conductivity, σ. Materials with a large amount of loss inhibit the propagation of electromagnetic waves. In this case, generally when $\frac{\sigma}{\omega\varepsilon} >> 1$, we consider the material to be a good conductor. Dielectrics are associated with lossless or low-loss materials, where $\frac{\sigma}{\omega\varepsilon} << 1$. Those that do not fall under either limit are considered to be general media. A perfect dielectric is a material that has no conductivity, thus exhibiting only a displacement current. Therefore it stores and returns electrical energy as if it were an ideal capacitor.

In the case of lossy medium, i.e. when the conduction current is not negligible, the total current density flowing is:

Where, σ is the conductivity of the medium; ε' is the real part of the permittivity & ε is complex permittivity.

The size of the displacement current is dependent on the frequency ω of the applied field E; there is no displacement current in a constant field.

In this formalism, the complex permittivity is defined as:

$$\varepsilon = \varepsilon' + i\frac{\sigma}{\omega} \text{or } \varepsilon = \varepsilon' + i\varepsilon''$$

In general, the absorption of electromagnetic energy by dielectrics is covered by a few different mechanisms that influence the shape of the permittivity as a function of frequency:

Relaxation effects associated with permanent and induced molecular dipoles. At low frequencies the field changes slowly enough to allow dipoles to reach equilibrium before the field has measurably changed. For frequencies at which dipole orientations cannot follow the applied field due to the viscosity of the medium, absorption of the field's energy leads to energy dissipation. The mechanism of dipoles relaxing is called dielectric relaxation and for ideal dipoles is described by classic Debye relaxation.

Resonance effects, which arise from the rotations or vibrations of atoms, ions, or electrons. These processes are observed in the neighborhood of their characteristic absorption frequencies.

Interaction of EM Waves

In a material medium the character, like wave impedance wave velocity etc. of an EM wave differs from that in free space. The material media influence the field quantities by way of electrical polarization or magnetization.

The dielectric materials are sensitive to electric field. Electric field (E) and displacement flux (D) are related to material electric polarization P by

$$D = \varepsilon_0 E + P \tag{30}$$

where $\varepsilon_0 = 8.86 \times 10^{-12}$ Farad is the permittivity of free space. Usually material polarization P is proportional to electric field i.e. $P = \chi_e E$, where the proportionality constant χ is called the electric susceptibility, and then

$$D = \varepsilon_0 E| + \chi_e E = \varepsilon_0 E\left[1 + \left(\chi_e - \varepsilon_0\right)\right] = \varepsilon_r \varepsilon_0 E \tag{31}$$

Where $$\varepsilon_r = 1 + \left(\chi_e - \varepsilon_0\right) \tag{32}$$

This is the relative permittivity of the material medium ε_r, which is a non dimensional parameter.

Magnetic materials are influenced by the EM wave magnetic component. Magnetic field (H) and flux induction (B) are related by the material's magnetization (M) by [21]:

$$B = \mu_0 (H + M) \tag{33}$$

Where $\mu_0 = 0.4\pi \times 10^{-6}$ Henry per meter is the permeability free space.

In a limiting case, M can be taken proportional to the magnetic field i.e. $M = \chi_m H$, Where χ_m is the magnetic susceptibility and then we have

$$B = (1 + \chi_m)\mu_0 H = \mu_r \mu_0 H \tag{34}$$

Where $$\mu_r = (1 + \chi_m) \tag{35}$$

This equation defines the relative magnetic permeability of the material medium μ_r, it is also non- dimensional.

The quantities ε_r and μ_r define the intrinsic impedance Z and the refractive index n of the material medium as

$$n = (\varepsilon_r \mu_r)^{\frac{1}{2}} \tag{36}$$

$$Z = Z_0 \left(\frac{\mu_r}{\varepsilon_r} \right)^{\frac{1}{2}} \tag{37}$$

where $Z_0 = \left(\frac{\mu_0}{\varepsilon_0} \right)^{\frac{1}{2}} = 377\,\Omega$ is the impedance of the free space.

The materials which have the complex indices of refraction are called lossy materials. The index of refraction includes electrical as well as magnetic effects, the imaginary part accounts for losses. At microwave frequencies, the loss is due to the finite conductivity of the material, as well as a kind of molecular friction experienced by molecules in attempting to follow the alternating fields of an impinged wave during polarization or magnetization process in the material. Instead of n, it is customary to express ε_r and μ_r, is normally given as:

$$\varepsilon_r = \varepsilon_{r'} - j\varepsilon_{r''} \tag{38}$$

$$\mu_r = \mu_{r'} - j\mu_{r'} \tag{39}$$

Where real part of the relative permittivity or permeability single prime is a measure of the extent to which the material will be polarized or magnetized by the application an electric magnetic field respectively. The imaginary part denoted by double prime is a measure of the energy losses incurred in re-arranging the alignment of the electric or magnetic dipoles as according to the applied ac fields. For the real part, polarization/ magnetization is in phase with the applied alternating E /H field that is maxima and minima of E/H coincide with P/ B where as lossy part, polarization/ magnetization is out of phase with the applied alternating E/H i.e. E/H are displaced by 90° with P/B. Further, if an ac electric field is applied across dielectric material, an ac displacement current is observed, which is due to the oscillation of electric dipoles with the field. Therefore, the dielectric materials posses an ac conductivity apart from dc conductivity due the migration of free charge carriers. Then the total contribution to dielectric loss is expressed as

$$\varepsilon_{r''} = \{(\sigma\mathrm{dc}/\omega\varepsilon_0) + \varepsilon_{ac''}\} \quad (40)$$

Where σdc is the conductivity, ω is angular frequency; ε_0 is the permittivity of free space and ε''_{ac} is loss contribution due to ac conductivity at high frequencies. Although ε''_r & μ'' show the lossy behaviour in the material, but it is also customary to express the losses by loss tangents. The dielectric and magnetic loss tangents are given by

$$\tan\delta_m = \varepsilon''_r / \varepsilon'_r \quad (41)$$

$$\tan\delta_m = \mu''_r / \mu'_r \quad (42)$$

Types of EM Wave Absoeber Preparation

Microwave lossy materials absorb the EM wave in microwave frequency regions and convert it into heat. But there is no such material which can absorb the complete frequency range (1-40GHz). Materials used for lower microwave ranges 1-3GHz and 3-8GHz are not same those used for middle ranges 8-12GHz. 8-18GHz; or used for higher ranges18-27GHz and 27-40GHz. Thus, materials must be chosen with respect to a single frequency or a particular frequency range to be absorbed.

Materials used for absorbing microwaves are mainly two types: (*i*) dielectric or (*ii*) magnetic. As a host, binder material are always dielectric. If absorbing filler is also dielectric, and the absorption occurs due to dielectric or resistive losses then it is known as dielectric absorber. This type of absorber are characterized by complex permittivity $(\varepsilon_r = \varepsilon'_r - j\varepsilon'')_r$ and complex permeability ($\mu_r = \mu_r' - j\mu_r''$) components are essentially $\mu_r' = 1$and

$\mu_r'' = 0$, same as for free space. On the other hand the absorbers that are designed employing magnetic fillers are known as magnetic absorber. The complex permeabilities for this category are different than that of free space, but the magnetic materials available for use have in general permitivities higher than their permeabilities. Therefore, the magnetic absorbers show both magnetic and dielectric loss properties and they are characterized by both the complex permitivities and complex permeabilities.

The EM requirements for absorbing material demand that the dielectric and magnetic properties of the material should provide a specific impedance profile to the incident wave so that the reflections at the material surface are minimized, and there should be some mechanism for EM absorption within the material. Such requirements are generally met by the composite absorbers. Composite materials are composed of a minimum of two phases: (i) a binder matrix that can be a polymeric thermoplastic, thermosetting resins; various kind of rubber compounds and some ceramic materials for special applications; and (ii) absorbing fillers that can be magnetic or dielectric materials such as different kind of ferrites, carbonyl iron, metal powder/flakes, conducting polymers, graphite/carbon powders etc.. High dielectric constant materials such as TiO_2 or $BaTiO_3$ are also suitably used to reduce the thickness of the absorber. Apart from these, sintered ferrite tiles and components are also used as microwave absorbers in specific applications. Some of the filler materials are discussed in this section.

Conducting Carbons

Use of conducting carbon black in absorber preparation is well known. Conducting carbons in various forms such as acetylene black, graphite powder/flake, short carbon fibers etc. are employed in preparation of microwave absorbers. The carbon materials because of its high conductivity reflects the EM wave and as such employed in EMI shielding working on the reflection principle. The ultra fine conducting carbon particles suspended in liquid such as water and alcohol, together with a small amount of a polymeric binder is painted on surfaces be shielded. The liquid carrier evaporates, thus allowing the carbon particles to be essentially in direct contact. Relatively, newer carbon material exfoliated graphite flakes (called worms) can be converted into flexible EMI gaskets by compression molding without use of any binder material. During exfoliation, graphite compound with some foreign species under special treatment forms intercalates between the graphite layers and material expands typically by over 100 times along the c-axis, resulting a flake or worm like structure. Compression of the resulting worms causes the worms to be mechanically interlocked to one another, so that a sheet is formed without a binder. When a carbon material is to be used as the EM wave absorbers, it is dispersed in to the insulating matrix, resulting into an absorbing flat sheets or pyramidal foams. It should bear appropriate values of e_r' & e''_r or a proper tan δ_ε values for a target frequency range.

Conducting Polymers

Organic conducting polymers have emerged as a new class of electro-active materials. Although these were known for several decades, but attracted much attention, when it was reported that poly acetylene treated with iodine shows high conductivity values up to 10^3 S/cm. Since then the research on fundamental and technological issues of conducting polymers have been intensified to explore their potential applications for EMI shielding, microwave absorption, antistatic coatings, rechargeable batteries, electro-chromic displays, light emitting diodes, solar energy conversion and other technological systems. Poly acetylene is not the only conducting polymer, many other polymers such as polypyrrole, polyaniline, polythoiphene, polyphenylene, poly phenylene-vinylene etc., having conjugated double in their chemical structure become conducting upon doping with an oxidizing or reducing agent. The double bonds are unsaturated bonds, and can be easily attacked by chemical methods. Oxidation removes electrons from the polymer and free sites (holes) are created, into which other electrons by hopping mechanism can move so that finally the whole system in the conjugated bonds becomes mobile and hence conducting. Upon reduction, additional electrons are transferred to the polymer. Conduction mechanism in these polymers is yet not clear, but it has been proposed that non linear excitations such as solitons and polarons play an important role in the transport of electrical charge. Among the whole conducting polymer family, polyaniline is more extensively used because of its high stability, flexibility in synthesis, process ability, high yield and efficiency. Polyaniline exists in a variety of forms depending upon its oxidation states. The most common forms are known as emeraldine salt and emeraldine base form. These two differ in chemical and physical properties. The green protonated emeraldine salt form as synthesized, has conductivity on a semiconductor level of the order of 10^0 S/cm, many order of magnitude higher than that of common polymer ($< 10^{-9}$ S/ cm) but lower than that of typical metals ($> 10^4$ S/cm). The protonated salt form e.g. polyaniline hydrochloride converts to insulating blue emeraldine base form when treated with ammonium hydroxide.

Ferrite Materials

Ferrite is commonly used as a generic term for describing a class of magnetic ceramics which are composed of iron oxide as a major component and other transition metal ions. It should not be confused with the metallurgical term used for the compound Fe_3C (ferrite phase) in the metallurgy of iron. Ferrites exist in various crystal structures, and exhibit different magnetic and dielectric properties depending upon the arrangement of oxygen anions around metal cations, forming tetrahedral, octahedral, dodecahedral and bi-pyramidal oxygen coordination around the cations. The magnetic properties of ferrites derive directly from the electron configuration of the cations and their interaction with each other. The magnetic cations in the ferrite form two sublattices. The magnetic ordering inside of a sublattice is usually collinear ferromagnetic. But the magnetic interaction between magnetic cations in different sublattices is through intermediate oxygen ion, termed as super exchange interaction. The nature of super exchange interaction depends not only on the type of magnetic ions, but strongly depends on the bond length and bonding angle. The sublattices are antiferromagnetically coupled, and presence of different types & number of magnetic ions in different lattices decides the resultant magnetic moment, giving rise to ferrimagnetism in the ferrites. Like ferromagnetic materials, ferrites also show spontaneous magnetization i.e. magnetic induction in absence of magnetic field. Ease of preparation, remarkable flexibility in tailoring the properties and low cost make ferrites the first choice in the search of materials for microwave device as well as absorber applications. Structurally they are grouped in following four categories.

(*i*) Spinel Ferrites

(*ii*) Hexaferrites

(*iii*) Magnetic Rare Earth Garnet

(*iv*) Perovskite or Orthoferrites

Out of these four categories, only first one i.e. spinels ferrites are extensively studied for microwave absorber applications. Spinels are generally employed for absorber application in lower microwave frequency ranges up to 3 GHz. Hexaferrites on the other hand can be employed for absorber application in the higher microwave frequency range from 2–40 GHz and beyond. These two will be discussed in detail.

Spinel Ferrites

The spinel structure is derived from the naturally occurred a nonmagnetic mineral spinel $MgAl_2O_4$ [52]. Analogous to this, magnetic spinel has the general compositional formula $M^{2+}Fe^{3+}_2O_4$ where trivalent Al is replaced by Fe^{3+} and M is the divalent metal ions. M is usually replaced by Mn, Fe, Co, Ni, Zn, Cu, Mg, Cd, Li etc. or more often a combination of these to have tailor made properties in the ferrite. When M is substituted by Fe^{2+}, magnetite Fe_3O_4 is formed, which is also a naturally occurring well known magnetic oxide since ancient times. The eight formula unit cell (smallest crystallographic repeat unit) of a spinel ferrite is composed of a cubic close packed arrangement of 32 oxygen ions with 64 tetrahedral interstitial sites (8 of which are occupied) and 32 octahedral interstitial sites (16 of which are occupied). In a normal spinel all of the divalent cations reside on the tetrahedral sites e. g. zinc ferrite $Zn^{2+}[Fe^{3+}_2]O_4$. In the inverse spinel all of the divalent cations reside on the octahedral sites e.g. nickel ferrite Fe^{3+} $[Ni^{2+}Fe^{3+}]$ O_4. In mixed spinel these divalent cations are distributed between both the sites e.g. Ni-Zn ferrite Zn^{2+}_{δ} $Fe^{3+}_{1-\delta}$ $[Ni^{2+}_{1-\delta}$ $Fe^{3+}_{1+\delta}]$ O_4, $0<\delta<1$. Addition of Zn in inverse ferrite increases the magnetization. At microwave frequencies, the magnetization occur due to the moments of the spin electrons, domain wall motion and rotation are not operative. If an electron is placed into a bias magnetic field, its magnetic moment aligns with the field to minimize its potential energy. If a microwave field is applied perpendicular to the bias field, magnetization will precess around the equilibrium direction with the frequency of the microwave field. In real materials there are damping forces, opposing the precessional motion, and magnetization relaxes back to the steady states. The same phenomena apply collectively to the magnetization vector (all spin aligned to one direction). If the frequency of the applied microwave field (i.e. EM wave) matches with that frequency of precessional motion of magnetization vector, ferromagnetic resonance (FMR) occurs. At resonance the interaction between the magnetization and the microwave field is very strong, and the EM energy gets absorbed. Energy stored in the process of magnetization is expressed as μ'_r, while the losses during relaxation process as μ''_r. μ''_r reaches a maximum value at the resonance frequency (f_r). Higher μ''_r corresponds with better absorption due to larger energy loss. Spinel ferrite with a higher μ''_r, however, tends to resonate at a lower frequency, and this has prevented these materials from being used at a higher GHz frequency range. This was first observed by J. L. Snoek. He found that the product of f_r and μ''_r is constant. This is expressed as:

$$f_r(\mu_{r'} - 1) = \left(\frac{4}{3}\right)\gamma M_s \tag{43}$$

where M_s is the saturation magnetization per cm^3 and γ is the gyro-magnetic constant. It is clear from eqn. (43) that f_r can be increased if $\mu_{r'}$ is decreased and M_s is increased. The highest possible M_s is achievable in case of Mn-Zn ferrite. Table 1 shows the values of $\mu_{r'}$ & $\mu_{r''}$ for Ni-Zn and Mn-Zn spinels ferrites in the frequency range of 3 MHz to 3 GHz. Spinel ferrites are generally employed for absorber applications in lower GHz region up to 3 GHz.

Hexaferrite means a group of ferromagnetic oxides possessing crystal structures closely related to hexagonal closed packing. Hexaferrites are an important class of materials because of their use in higher GHz frequency region. Spinel ferrites with their cubic structures, as discussed earlier, show isotropic behaviour, and need external bias field in excess of 20 kG to achieve ferromagnetic resonance condition. Contrary to it, hexaferrite do not require external bias field in microwave operation, because of their intrinsic high anisotropy field (H_A), they are said to be self biased. There are several types of hexaferrites designated as M, Y, W, X, Z and U. M type shows very high uniaxial anisotropy field along the c-axis, while other members show a moderate in plane anisotropy on the basal plane of their hexagonal structure. The crystal and magnetic structure of the different types of hexaferrites is quite complex, but all types are inter-related.

Among the hexaferrites family, M type $BaFe_{12}O_{19}$ is well known because of their tonnage requirement in low cost permanent magnets. The substituted Ba-hexaferrites have also been extensively studied in high frequency absorber applications. U type hexaferrite $Ba_4Me_2Fe_{36}O_{60}$ is least explored member of the hexaferrite family.

Fabrication Geometries For Absorber

Both types of absorbers, weather dielectric or magnetic can be fabricated in single layer with homogenous dielectric/magnetic properties through out the sample; or in multilayer with homogeneous/graded dielectric/magnetic properties in different layers; or with geometrical transition, depending up on their single, multiple or broadband frequency application. Conceptually, absorbers can be divided into two broad categories: (*i*) graded impedance type absorber, and (*ii*) resonant absorber based on their geometries.

Table 1: Spectral Complex Permeability Data of Ni-Zn and Mn-Zn Ferrites

Frequency in MHz	Ni-Zn Ferrite		Mn-Zn Ferrite	
	μ_r'	μ_r''	μ_r'	μ_r''
3	963	909	508	536
10	259	557	201	277
30	46.7	233	90.5	155
100	6.94	75.5	24.0	73.4
200	2.20	39.2	7	41.7
300	0.89	26.5	2.75	28.4
600	0.59	13.3	0.48	14.1
1000	0.80	7.92	0.02	8.36
1500	0.89	5.86	0.16	5.36
2000	0.98	3.79	0.18	3.39
2500	0.99	2.43	0.5	2.68
3000	1	1.07	0.6	2.16

Graded Impedance Type Absorber

These are designed by varying the impedance from that of free space (377 Ω) at its front surface to lower impedance, more lossy material at its rear surface. Incident EM wave finds free space as it enters the absorber at the front surface, and is attenuated as it propagates through the absorber medium which becomes more and more lossy towards rear surface. Instead of having absorber of uniform properties, the dielectric constant is adjusted to be very low at the front surface and increases to a relatively high number at the rear. Consequently, the front face reflection and the overall thickness, which is required to make the back face reflection negligible, are greatly reduced. Extremely low reflection levels <50 dB can be achieved in this manner, but appreciable absorber thickness is required, a few meters to a few cm depending upon the frequency region 100 MHz to 100 GHz. These absorbers can further be divided into two categories:

(*a*) geometrically shaped absorber, and (*b*) graded interface absorber

Geometrically Shaped Absorber

Here the impedance gradient is achieved by geometrical shaping of a medium of constant impedance (Fig. 4). This provides the opportunity for a much more complete transition to take place from free space to the dissipative medium. Geometrical shaping means, the breaking of the front surface into an aggregate of shaped pointed elements such as pyramids, cones, sinusoidal or triangular, where the axis of individual elements is oriented perpendicular to the plane of the absorber. The EM wave entering such an absorber medium encounters a smoothly changing ratio of medium to the adjacent free space. This is similar to the case where the actual properties are changing smoothly. When these elements are sharp, any interface reflection is negligible and only a reflection remains that is directly related to the gradient or impedance change per wave length of depth into the medium. Such geometrical transition, however, requires large physical thickness. Therefore, carbon loaded foam based pyramidal absorbers as shown in the figure, are utilized mainly in anechoic chambers. These types of absorbers can provide a reflection level < -50 dB, but thickness requirement may be in excess of 10 λ; e.g. at 10 GHz, $\lambda = 3$ cm, and absorber thickness needed in excess of 30 cm or 1 ft.

Graded interface absorber

These are flat broadband absorber made up of stacked layers (Fig. 5). Each layer consists of homogenous materials, giving one type of EM properties. But different layers have different EM properties. On stacking the layers, there exists a gradual change in impedance from front to rear layer; or an impedance gradient between the layer's interfaces. The thin layers of carbon loaded foams with increasing order electrical conductivity from front to rear side can be stacked to realize a practical thin flexible and broadband absorber. Another way to realize graded interface absorber is that instead of stacking the each layer with uniform impregnation, a tapered impregnation with a continuous variation in electrical properties is done. They can provide reflectivity better than -20 dB over the appreciable frequency range, and are only about $\lambda/4$ thick at low design frequency limits. The performance of these absorbers is limited by the front-face reflections resulting from the relatively high dielectric constant of the material of construction. A magnetic absorber with graded interfaces is also realizable by stacking the ferrite layers of different EM properties. Absorbers of this category can be employed around antennas to improve side-lobe control.

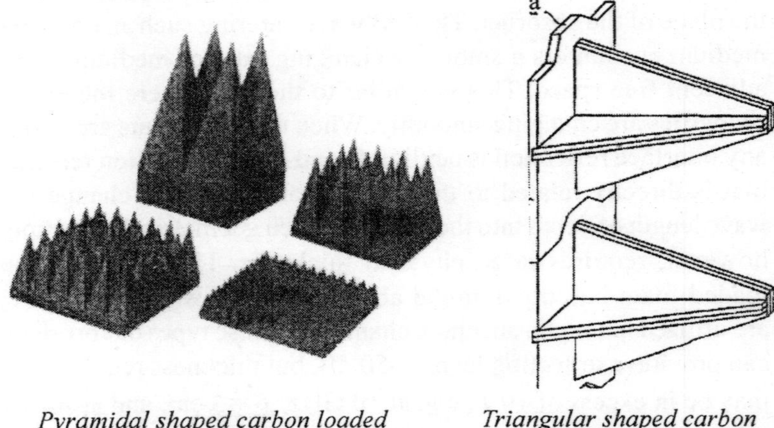

Pyramidal shaped carbon loaded foam absorber | *Triangular shaped carbon ferrite absorber*

***Fig. 4:** Typical geometrically shaped absorber used in anechoic chambers*

Resonant absorbers may be thought of as circumventing the problem of high reflections at air – absorber interface, by canceling this reflection with another reflection from the rear absorber surface, which is directly attached to a metal conductor. Resonant absorbers derive their name from the fact that the conditions for reduction of reflected EM wave are satisfied, in general, only at one or more discrete resonance frequencies. It may be shown that these two vectors cancel at a frequency where the thickness of the absorber is essentially a quarter-wave or odd multiples of a quarter-wave. The well known Salisbury screen, Jaumann and Dallenbach layer absorbers belong to this category.

Salisbury Screen and Jaumann Absorbers

The Salisbury screen is created by placing a thin resistive sheet on a low dielectric constant spacer in front of a metal plate (Fig. 6, Fig. 7). Here the resistivity of the sheet must equal 377 ohms per square ($\Omega/\square$), which is equal to the impedance of free space. Typically, a foam or honeycomb spacers are used to have a low dielectric constant in the range 1.03 to 1.1. The spacer thickness must be equal to the electrical quarter wavelength or odd multiple of it. This absorber has the analogy in transmission line theory. A simple transmission line termination can be effected by placing a pure resistance across a shorted transmission line at a point the quarter wavelength apart in front of the short. There is complete absorption at that frequency, when this resistance is equal to the characteristic impedance of the line.

The Single layer Salisbury screens are narrow band absorber. It can show a reflection level < -40 dB in approximate thickness of 7.5 mm at 10 GHz (λ = 30 mm), but the bandwidth associated with them are not sufficient to render any fruitful application in present scenario of multi band and high agility radar systems.

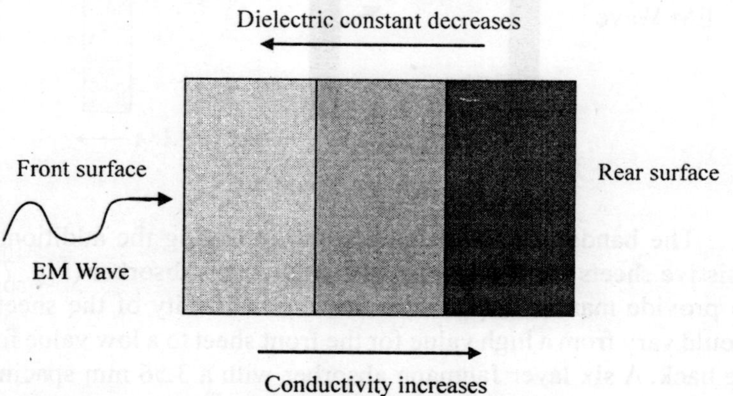

Fig. 5: Graded interface absorber

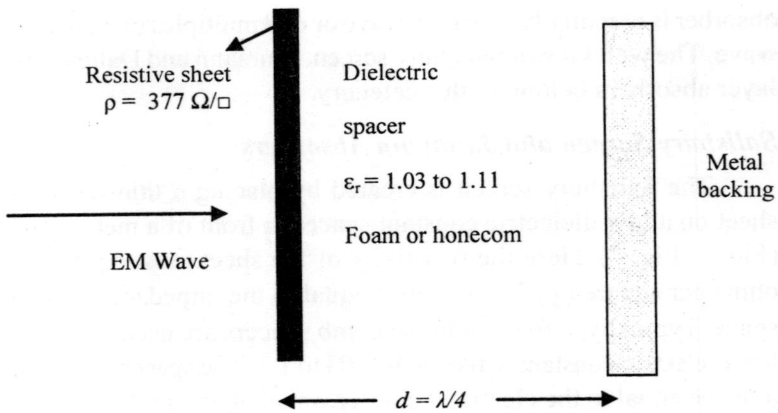

Fig. 6: Salisbuary screen absorber

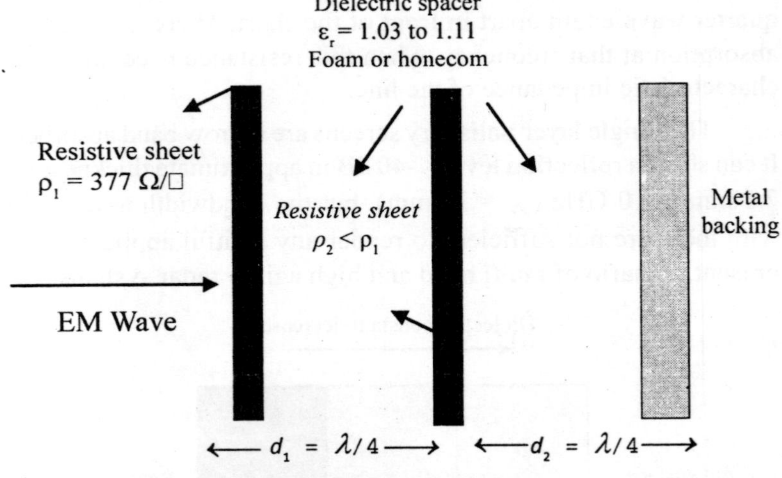

Fig. 7: Jaumann absorber

The bandwidth can be improved by adding the additional resistive sheets and spacers to form a Jaumann absorber (Fig. 7). To provide maximum performance, the resistivity of the sheets should vary from a high value for the front sheet to a low value for the back. A six layer Jaumann absorber with a 3.56 mm spacing between layers with a spacer of $\varepsilon_r = 1.03$ provide an average RCS reduction of -30 dB over the frequency range of 7 – 15 GHz. These types of absorbers are employed for making radar absorbing structural (RAS) composites.

Dallenbach Layer Absorber

The Dallenbach layer absorber consists of a homogeneous lossy layer of dielectric/ magnetic materials backed by a metal plate. The reflection at the surface of the material is due to the impedance change seen by the wave at the interface between the two media. The materials dielectric and magnetic properties are so adjusted that the condition for low reflection at the interface is created. In this case the attenuation will depend on the loss properties of the material (ε_r'' , μ_r'') and the electrical thickness. This category of absorbers are much thinner than other types, they are preferred for most applications where wide operational band width is not a requirement. They are customarily chosen to improve antenna patterns by covering parts of antennas and to improve radar performance by covering nearby reflecting surfaces. Their physical flexibility is often useful in coating surfaces of complex shape. They are usually mounted directly on the metal that is causing the reflection. Where the surface is sharply curved or small in terms of wave lengths, some reduction in performance is to be expected. For cases where the material cannot be mounted directly on metal, it is necessary to provide a conducting rear surface (such as foil or silver paint) to achieve low reflection at the resonant frequency. This category of absorber can be realized in the form of thin, flexible, elastomeric sheets, or spray able paints. When mounted directly on a metal surface, the reflectivity is generally better than -15 dB with 15% bandwidth around resonance. The absorbers prepared in the present work, belong to Dallenbach layer, whose principle of operation has been described in the next section.

Principle of Operation of Dallenbach Layer Absorber

A typical absorber consists of an absorptive layer that is backed by a metal sheet or oil, as shown in Fig. 8. When the microwave impinges on the absorptive layer, electrons present in the materials move under the microwave-induced magnetic field, which then induces eddy current loss due to resistive heating. These eddy currents create secondary small magnetic fields opposite to the excitation field causing reflection of the incident microwave. In this process, a significant part of the incident wave is reflected at the air – absorber interface, some get absorbed in the material due to dielectric and magnetic losses (ε_r'' , μ_r'') and rest is reflected back from the metal backing. These two reflected waves are out of phase by 180^0 and cancel each other at air-absorber interface, if the complex permittivity, permeability and thickness of the absorbing layer are adjusted according to the quarter wave thickness (d) criteria

$$d = \lambda_0 \Big/ 4(|\mu_r|.|\varepsilon_r|)^{\frac{1}{2}} \quad (44)$$

where λ_0 is the free space wavelength of incident wave and $|\mu_r|$ & $|\varepsilon_r|$ are the moduli of μ_r & ε_r .

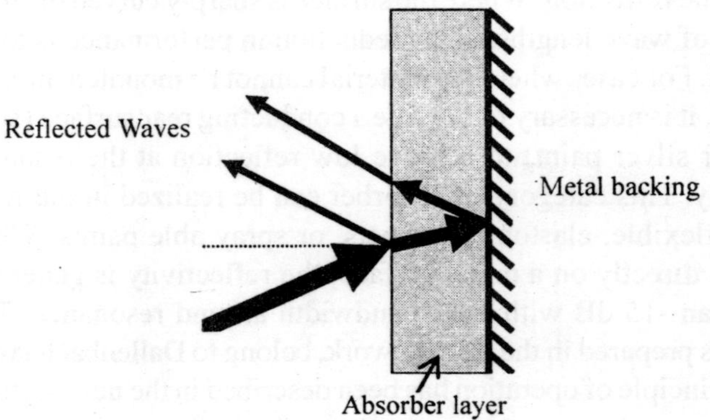

Fig. 8: Scheme of a single layer absorber

In the case of metal backed single-layered absorber, the normalized input impedance with respect to its impedance in free space (Z), and reflection loss (R_L) with respect to the normal incident plane wave are given by:

$$z = (\mu_{\downarrow} r / \varepsilon_{\downarrow} r) \uparrow (1/2) \ \tanh[(-j2(2c)((\mu_{\downarrow} r . \varepsilon_{\downarrow} r) \uparrow (1/2) f . d] \quad (45)$$

$$R_L = -20\log_{10}\left[\left|\frac{(z-1)}{(z+1)}\right|\right] \quad (46)$$

where $\mu_r = \mu_{r'} - j\mu_r''$, and $\varepsilon_r = \varepsilon_{r'} - j\varepsilon_r''$, are the relative complex permeability and permittivity of the absorber medium respectively. d is the sample thickness. c is the velocity of light and f is the frequency of microwave in free space.

The above equations (45) and (46) look relatively simple, but the relation between the material properties and the minimum reflection is not so simple. The high loss tangent in the material alone is not indicative of minimum reflection. In the minimization problem one has to optimize the properties for two loosely connected phenomena viz. the reflection from the front surface (impedance matching) and the absorption inside the absorber layer. The impedance matching condition representing the perfectly absorbing properties is given by Z = 1. This demands the optimization of the six parameters $\mu_r', \mu_r'', \varepsilon_r', \varepsilon_r''$, f, and. d

Samples Preparation

For the experiment, samples are prepared by using absorbing pigments and commercially available binder, two components poly urethane 50% Polyol and 50% Hexamethylene Diisocynate. The mixture of pigment and binder were thoroughly mixed in Mortar and pestle. Wet mixing is used by taking the Acetone medium to homogenize the mixture. After rigorously mixing the pigment and binder it pored in the rectangular dye. The dyes were prepared of wood of size 15mm X30mm. Magneese- zinc ferrite and nickel-zinc ferrite were chosen as pigment and large number of samples were prepared by using these pigments of different proportions by keeping binder constant. The ratio of pigments was varied from 10% to 80%.

The permittivity, permeability and reflection loss were measured by Agilent vector network analyzer.

Microwave Measurements

Early works in the field of microwave measurements on materials can be found in the book by Von Hippel and M. Sucher. These books remain the reference works in the measuring dielectric permittivities and magnetic permeabilities of homogeneous materials. Methods used at that time were limited in frequency band measurements, permitting only the determination of complex permittivity and permeability at fixed frequencies. One such popular method is known as cavity perturbation technique, also used, initially, in the present work for microwave measurements on dielectric absorbers. W. B. Weir first presented the broadband microwave measurement method based on the reflection and transmission coefficients (S-parameters) resulting when a test sample was inserted into a waveguide or a TEM transmission line. From measured S-parameters, complex permittivity and permeability could be determined in the range of 50 MHz to 18 GHz. In 1987 Hewlett Packard published a product note related to measuring the complex permittivity and permeability with HP network analyzer. This method is currently in use and also utilized for most of the microwave measurements of the present work.

S-Parameter Technique

With the advent of computer controlled network analyzer and advanced automated computational aids, presently, this is the most widely used technique to determine the complex permittivity and permeability of the materials. In this technique, network analyzer measures the magnitude and phase response of a device/material by comparing the incident signal with the signal transmitted by the device/material or reflected from its inputs as shown in the Fig. 9. The measured response is expressed in terms of complex scattering parameters (S_{11} or S_{22} for reflection and S_{21} or S_{12} for transmission). The basic theory relates the S_{11} and S_{21} to the actual reflection coefficient (Γ) and transmission coefficient (T) based on the assumption made that there is no source and load

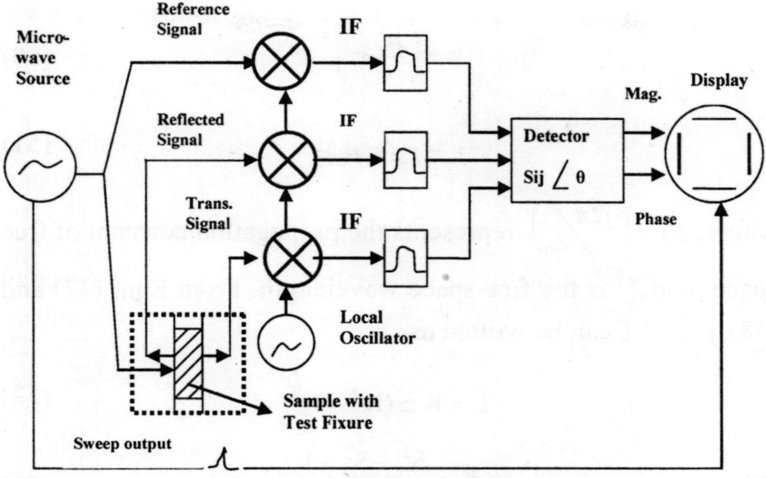

Fig. 9: Block diagram of Network Analyzer

mismatches, no air gap between material and the conducting walls of wave-guide sample holder, and the surfaces of the sample are planar and normal to the direction of wave propagation.

$$S_{11} = \frac{\Gamma(1-T^2)}{1-\Gamma^2T^2} \quad (47)$$

$$S_{21} = \frac{T(1-\Gamma^2)}{1-\Gamma^2T^2} \quad (48)$$

However, actual reflection coefficient (Γ) at the air-sample interface, and transmission coefficient (T) are given by

$$\Gamma = \frac{Z-1}{Z+1} \text{ and } T = e^{-\gamma d} \quad (49)$$

where, d is the sample thickness, Z and γ are the normalized characteristic impedance and propagation constant of the sample material. These are related to the complex permittivity, $\varepsilon_r = \varepsilon_r' - j\varepsilon_r''$ and complex permeability, $\mu_r = \mu_r' - j\mu_r''$ relative to free space by the following relationship:

$$z = \left(\frac{\mu_r}{\varepsilon_r}\right)^{\frac{1}{2}} \qquad (50)$$

$$\gamma = \gamma_0(\varepsilon_r \mu_r)^{\frac{1}{2}} \qquad (51)$$

where, $\gamma_0 = \left(\frac{j2\pi}{\lambda_0}\right)$ represents the propagation constant of free space, and λ_0 is the free-space wavelength. From Eqn. (47) and (48), Γ and T can be written as

$$\Gamma = K \pm (K^2 - 1)^{\frac{1}{2}} \qquad (52)$$

Where,
$$K = \frac{S_{11}^2 - S_2^2 + 1}{2S_{11}} \qquad (53)$$

$$T = \frac{S_{11} + S_{21} - \Gamma}{1 - \Gamma(S_{11} + S_{21})} \qquad (54)$$

In Eqn. (52), the $\pm$ are chosen such that $[\Gamma] < 1$. From (49), complex propagation constant, γ can be written as

$$\gamma = \frac{\left[\log_e \left(\frac{1}{T}\right)\right]}{d} \qquad (55)$$

From Eqn. (49) and Eqn. (50)

$$\frac{\mu_r}{\varepsilon_r} = \left[\frac{1+\Gamma}{1-\Gamma}\right]^2 \qquad (56)$$

$$\mu_r \varepsilon_r = \left(\frac{\gamma}{\gamma_0}\right)^2 \qquad (57)$$

Dividing Eqn. (57) by Eqn. (56) and multiplying them, we obtain

$$\varepsilon_r = \frac{\gamma}{\gamma_0}\left(\frac{1-\Gamma}{1+\Gamma}\right) \qquad (58)$$

$$\mu_r = \frac{\gamma}{\gamma_0}\left(\frac{1+\Gamma}{1-\Gamma}\right) \tag{59}$$

From the basic theory given above, ε_r and μ_r can be determined by measuring the S_{11} and S_{21} parameters.

The samples were shaped to fit exactly into a rectangular X-band wave-guide (WR90) of dimensions: 10.2 mm × 22.8 mm for the measurement of S-parameters. The complex scattering parameters that correspond to the reflection (S_{11} or S_{22}) and transmission (S_{21} or S_{12}) in the samples were measured using a vector network analyzer (HP/Agilent, PNA E8364B) in the X–band frequency range of 8.2 – 12.4 GHz. Full two port calibration was initially done on the test setup in both the forward and reverse directions, before commencing any measurements, to avoid any kind of errors due to the source match, load match, directivity, isolation and frequency response etc.. Agilent software module 85071 (version 'E') based on the theory given above was used to determine the complex permittivity and permeability from the measured scattering parameters. The module is standard, hardware locked software supplied by the HP-Agilent along with the latest version of their vector network analyzer (model PNA E8364B) for determination of complex permittivity and complex permeability of a solid type dielectric/magnetic materials exhibiting high loss.

As a general precaution, samples were made with exact dimensions for slide fitting into wave-guide sample holder, leaving no scope for error due to air gap between samples and conducting walls of the sample holder. Sample's surfaces were smoothed out to minimize the multiple reflections at the air – material interface. Same measurements were carried out for two sample thicknesses to avoid any ambiguity due to higher order modes excitation. Because of the high quality of the network analyzer, the instrumentation errors e.g., frequency instability, power variation of microwave signals, cables & waveguide connectivity etc. are assumed to negligible.

Reflection loss Measurement

For the measurement of reflection loss, samples were cut into a rectangular shape of size 10.2 mm × 22.9 mm to fit in rectangular wave-guide of X- band (WR90), same as for the measurement of S-parameters. The wave-guide fitted with sample was backed by a metal short. In some samples, voltage standing wave ratio (VSWR) was measured as a function of frequency using HP network analyzer (Model 6719 ES). The reflection loss (R_L) was then determined for each measured points using the expression: R_L (dB) = -20 log ç(VSWR–1)/(VSWR+ 1)ç. In other cases, R_L was directly measured from S_{11} parameter using the terminated one-port technique. In this case, R_L (dB) = - 20 $\log_{10}|S_{11}|$.

The reflection loss is a function of complex interplay of permittivity, permeability, absorber thickness and EM wave frequency. To get an ideal absorber, a variation problem with six variables i.e. real and imaginary permittivity $\left(\varepsilon_r', \varepsilon_r''\right)$; real and imaginary permeability $\left(\mu_r', \mu_r''\right)$; thickness (d) and frequency (f) should be solved, which is an extremely challenging problem. In the recent past some researchers have made great progress in design methods and techniques to predict these parameters to have an excellent absorber. However, still it is very difficult to find the dielectric and magnetic materials with the predicted parameters. Therefore, much of the designing is based on combination of existing materials and time-honoured method of trial and error. This latter approach is adopted in the present work. Variation of permittivity and permeability in the composite absorbers were obtained by varying the content of absorbing materials. Once the absorber is already prepared, the only variable parameters remained are d and f. The developed code utilized the measured values of ε'_r,ε''_r, μ'_r and μ''_r to determine the reflection loss in the composite samples based on a model of a single layered plane wave absorber proposed by Naito and Suetake, and finds out matching thickness (d_m) and a matching frequency (f_m), where minimum reflection loss occurs.

Results and Discussions

The work carried out may be summarized and concluded in the following manner: The electromagnetic (EM) and microwave absorption properties for different magnetic absorbers i.e. Mn-Zn ferrite and Ni-Zn ferrite polymer composites as observed are summarized in Table 2 & 3. In the all ferrite- polymer composites, the ferrite quantity is increased by weight from 30% to 80% successively and accordingly polyurethane content is taken.

Table 2: Summary of Results for Mn-Zn Ferrite –PU Composite Absorber

Mn-Zn ferrite based absorbers (Weight %)	*At ~10 GHz*				*Minimum* R_L *(dB)*	t_m *(mm)*	*Frequency (GHz)*
	ε_r'	ε_r''	μ_r'	μ_r''			
30	4.4	0.4	1.2	0.0	-27.5	22	11.7
40	4.9	0.5	0.9	0.2	-3.75	3.8	9.5
50	6.5	0.6	0.8	0.1	-6.2	3.0	12.2
60	9.0	1.0	1.0	0.0	-23.8	2.4	11.9
70	25.0	10.5	1.0	0.1	-15.5	2.2	9.9
80	52.5	24.0	0.0	0.0	-28.6	1.6	12.2

Table 3: Summary of Results for Ni-Zn Ferrite Based Magnetic absorber

Ni-Zn ferrite based absorbers (Weight %)	*At ~10 GHz*				*Minimum* R_L *(dB)*	t_m *(mm)*	*Frequency (GHz)*
	ε_r'	ε_r''	μ_r'	μ_r''			
30	5.1	0.6	1.0	0.1	-24.0	20.0	11.8
40	6.5	1.1	1.0	0.1	-19.0	10.0	9.2
50	8.7	2.0	1.0	0.1	-10.0	2.8	10.0
60	12.1	5.0	1.0	0.1	-23.0	2.4	9.4
70	16.5	6.0	1.1	0.0	-27.6	1.6	12.0
80	11.0	7.5	1.0	0.0	-17.2	2.0	12.0

Table 2 indicates that Mn-Zn ferrite composite (80% composition) has shown the best absorption properties with a minimum R_L = –28.6 dB at 12.2 GHz for a sample thickness of just 1.6 mm. This composition exhibited permittivity values, ε'_r = 52.5 and ε''_r = 24.0; and permeability values, μ'_r ~ 0.0 and μ''_r ~ 0.0 at 10 GHz.

The composite with 30% ferrite content has shown a minimum R_L= -27.5dB with a sample thickness of 22.0 mm at 11.7 GHz. At lower thickness, absorption of only 2 to 3 dB has been observed This composite exhibits permittivity values, ε'_r = 4.4 and ε''_r = 1.0; μ''_r ~ 0.0 and permeability values, μ'_r ~ 0.0 and μ''_r ~ 0.0 at 10 GHz. It is therefore concluded that for better absorption at lower thicknesses, higher ferrite content is needed.

Table 3 indicates that Nn-Zn ferrite composite (70% composition) has shown the best absorption properties with a minimum R_L = –28.6 dB at 12.0 GHz for a sample thickness of just 1.6 mm. This composition exhibited permittivity values, ε'_r = 16.5 and ε''_r = 16.0; and permeability values, μ'_r ~ 1.0 and μ''_r ~ 0.0 at 10 GHz.

The composite with 30% ferrite content has shown a minimum R_L= -24.0dB with a sample thickness of 22.0 mm at 11.8 GHz. At lower thickness, absorption of only -2 to -3 dB has been observed This composite exhibits permittivity values, ε_r' = 5.1 and ε_r'' = 0.6; and permeability values, μ_r' ~1.0 and μ_r'' ~ 0.0 at 10 GHz. It is therefore concluded that for better absorption at lower thicknesses, higher ferrite content is needed.

Future Prospects

Microwave absorbing composites were prepared and characterized for X-band frequency range (8.2 – 12.4 GHz). Work may be extended below and above the X-band microwave frequencies. A multilayer approach may be adapted for achieving wide band absorption absorption. As ferrites are quite heavy, therefore, requiring light weight microwave absorber. Thus, composite absorber may be prepared using the light weight materials like carbon nanotubes (CNTs), carbon microcoils (CMCs), polyaniline etc.

References

1. **Thomas C. Maloney, Nicola Bowler, Nathan L. Fischer,** Microwave Absorption Characteristics of Composites with Tungsten Coated Filler Particles, *IEEE Trans.* Vol.-15 no.-3, June 2008.
2. **M.Itoh, J.R.Liu, T. Horikawa, K.I. Machida,** Electromagnetic Wave Absorption Properties of Nanocomposite Powders Derived from Intermetallic Compounds and Amorphous Carbon, *Journal of Alloys & compounds*, 408-412, p.1400-1403; 2006.
3. **A.Das & S.K.Das,** *Microwave Engineering* (Tata MacDraw-Hill Publishing Co. Ltd. , New Delhi, India 2003) 445.
4. **T.Sue, K.Enokida & Y.Okawa,** *U.S.Patent* No. 6822541 B2, Nov. 23, (2004).
5. **R.A. Glbert**, *U.S. Patent* No. 6538596, Mar. 25, (2003).
6. **R.H. Victora,** *U.S. Patent* No. 6441771 B1, Aug. 27 (2002).
7. **K.Nakamura, H.Komori, M.Oda & K.Kanda,** *U.S. Patent* No. 5770304, Jun. 23 (1998).
8. **S.Y.Hong & H.R.Chang,** *U.S.Patent* No.5668070, Sep. 16 (1997).
9. **P.O.Chul Kim, Dai Gill Lee,** Composite Sandwitch Construction for Absorbing E Wave, *Composite structure* 87 (2009) 161-167.
10. **Nikawa, M.Chino, S.Kitahata & K.Kurata,** *US Patent* No. 5952953, Sep.14(1999).
11. **T.Yamane, S.Numata, T.Mizumoto & Y.Naito,** Electromagnetic Coompatibility, *IEEE International Symposium* on 19-23 Aug. 2002, 2(2002) 799.
12. **Nujiang Tang, Wei Zhong, Chktong Au, Yiyang Mangul Han, Kuanjiuh Lin and Youwel Du,** Synthesis and Microwave Absorption Properties of Carbon-nanocoils, *J.Phys.Chem.* C 2008,112, 19316-19323.
13. **W. H. Emerson,** *IEEE Trans.* Antenna and Propagation, AP-21 (4) (1973) 484.
14. **R. A. Stonier,** SAMPE Journal, Vol. 27 (4) (1991) 9.

15. **A. N. Yusoff, M. H. Abdulla, S. H. Ahmad, S. F. Jusoh, A. A. Mansor, S. A. A. Hamid,** *J. Appl. Phys.* 92 (2002) 876.

16. **H. Severin and J. P. Stoll, Z. Angew.** *Phys.* 23 (1967) 209.

17. **M. R. Meshram, N. K. Agarwal, B. Sinha and P. S. Misra,** *Magn. Magn. Mater.* 271 (2004) 207.

18. **A. Verma, R. G. Mendiratta, T. C. Goel and D.C. Dube,** *J. Electroceramics* 8 (2002) 203.

19. **P. Singh, V. K. Babbar, A. Razdan, R. K. Puri and T. C. Goel,** *J. Appl. Phys.* 87 (9) (2000) 4362.

20. **P. Singh, V. K. Babbar, A. Razdan, S. L. Srivastava and T. C. Goel,** *Mater. Sci. Engg.* B7 (2000)70.

21. **W. Phang, R. Daik and M. H. Abdullah,** *Thin Solid Films* 477 (2005) 125.

22. **S. Marchant, F. R. Jones, T. P. C. Wong and P. V. Wright,** *Synthetic Metal* 96 (1998) 35.

23. **S. K.Kwon, J.Moahn, G.H.Kim, C.H. Chun, J.S.Hwang and J.H.Lee,** *Polymer Engg. Sci.* 42 (2002) 2165.

24. **Y. Nikawa, M. Chino, S. Kitahata and K. Kurata,** *US Patent* No. 5952953, Sep. 14, (1999).

25. **K. Hatakeyama and T. Inui,** *IEEE Trans. Mag.*, 2(5) (1984) 1261.

26. **D.D.L. Chung,** *Carbon* 39 (2001) 279.

27. **L. Olmedo, P. Hourquebie and F. Jousse,** *Microwave Properties of Conductive Polymers* (Handbook of Organic Conductive Molecules and Polymers, Vol. 3, Edited by H. S.Nalwa; John, Wiley & Sons, Inc., New York, 1997) 367.

28. **C. K. Chiang, C. R. Fincher, Y. W. Park, A. J. Heeger, H. Shirakawa, E. J. Louis, S.C. Gau and A. G. MacDiarmid,** *Phys. Rev. Lett.* 39 (1977) 1098.

29. **R. Subramaniam and S. D. Deshpande,** *US Patent* 6900286 B2, May 31, (2005).

30. **H. Gao, T. Jiang, B. Han, Y. Wang, J. Du, Z. Liu and J. Zhang,** *Polymer* 45 (2004) 3017.

31. **H. Aghlara,** *Chinese Journal of Physics* 41 (2) (2003) 185.

32. **J. Deng, C. L. He, Y. Peng, J. Wang, X. Long, P. Lei and A. S. C. Chan,** *Synthetic Metal*, 139 (2003) 295.

33. **S. Koul, R. Chandra and S. K. Dhawan,** *Polymer* 41 (2000) 9305.

34. **P. Chandrasekhar and K. Naishadham,** *Synthetic Metals* 105 (1999) 115.

35. **P. Hourouebie, B. Blondel and S. Dhume,** *Synthetic Metal*, 85 (1997) 1437.

36. **L. Olmedo, P. Hourouebie, and F. Jousse,** *Synthetic Metal*, 69 (1995) 205.

37. **J. Stejskal and R. G. Gilbert,** *Pure Apll. Chem.* 74 (5) (2002) 857.

38. http:// www.nasatech.com/Briefs/Nov00/MSC22647.html

39. **A. Goldman,** *Modern Ferrite Technology* (Van Nostrand Reinhold,New York,1990)21

40. **M. P. Horvarth,** *Journal of Magnetism and Magnetic Materials*, 215 (2000) 171

41. **H. Ota, M. Kimura, R. Sato, K. Okayama, S. Kondo and M. Homma,** Electromagnetic Compatibility (1999) *IEEE International Symposium* on, 2 -6 Aug. 1999 2 (1999) 590.

42. **J. Smit and H.P.J. Wijn,** *Ferrites: Physical Properties of Ferrimagnetic Oxides in Relation to their Technical Applications*, (Philips Technical library,Netherland, 1959) 78 and 271.

43. **F. Mayer,** *US Patent* 5872534, Feb. 16, (1999).

44. **C. H. Peng, C. C. Hwang, J. Wan, J. S. Tsai, and S. Y. Chen,** *Mater Sci. Engg.* B, 117 (2005) 27.

45. **A. N. Yusoff and M. H. Abdulla,** *J Magn. Magn. Mater.* 269(2) (2004) 271.

46. **N. Matsushita, T. Nakamura and M. Abe,** *IEEE Trans.* Mag., 38(5) (2002) 3111.

47. **M. R. Meshram, N. K. Agarwal, B. Sinha and P. S. Misra,** *Magn. Magn. Mater.* 271 (2004) 207.

48. **T. Hayashi and Y. Sakai,** *US Patent* 6784419 Aug. 31, (2004).

49. **M. B. Amin and J. R. James,** *Radio and Electronic Engineer*, 51(5) (1981) 209.

50. **W. W. Salisbury,** *US Patent* 2599944, Jan. 10, (1952).

51. **R. H. Victora,** *US Patent* No. 6441771 B1, Aug. 27 (2002).

52. **V. B. Bregar, A. Znidarsic, D. Lisjac and M. Drofenic,** *Materiali in Tehnologije* 39 (2005) 3.

53. **M. Sucher and J. Fox,** *Hand Book of Microwave Measurements* (3rd ed., Polytech. Inst., Brooklyn, New York 1963).

54. **W. B. Weir,** *Proc. of the IEEE,* 62 (1) (1974) 33.

55. Hewlett-Packard, Microwave Network Analyzer Product Note, 8510-3 (1987).

56. **H. M. musal and H. T. Hahn,** *IEEE Trans. Magn.* 25 (1989) 3851.

57. **Y. Naito,** *Elec. Commn. in Japan* 52-B (1) (1969) 71.

58. **Y. Naito and K. Suetake,** *Elec. Commn. in Japan* 52-B (7) (1969) 61.

59. **Y. Kim, Y. C. Chung, T. W. Cang and H. C. Kim,** *IEEE Trans.* Mag. 12 (2) (1996) 555.

60. **S. Q. Zhang, C. G. Huang, Z. Y. Zhou and Z. Li,** *Mater. Sci. Engg.* B90 (2002) 38.

61. Stealth in Focus, Defence Attache No. 6 (1986) 27.

62. **B. Chambers and A. Tennant,** *Electronics Letters* 30 (18) (1994) 1530.

63. **J. Ajadmanjiri,** Structural and Electromagnetic Properties of Ni-Zn Ferrites Prepared by Sol-gel Combustion Method, *Materials Chemistry and Physics* 109 (2008) 109.

64. **Huang Aiping,He. Huahui, Feng Zekun, Wang Shilie,** Study on Electromagnetic Properties of Fe-poor Composition, *Materials Chemistry and Physics* 105 (2007) 303.

16

Neutrino Oscillations And Mass Hierarchies

K. Chaturvedi, M. Tripathi
Jai Prakash and Ashish Mautiyal

Neutrinos, they are very small. they have no change and have no mass And do not interact at all. The earth is just a silly ball To them, through which they simply pass Like dustmaids down a drafty hall Or photons through a sheet of glass They snub the most exquisite gas. Ignore the most substantial wall Cold –shoulder stell and sounding brass. Insult the stallion in his stall. And, scorning barriers of class. Infiltrate you and me ! like tall And painless guillotines into the grass At night, they enter at Nepal And pierce the lover and his lass From underneath the you call It wonderful, I call it crass.

The history of neutrino started in 1930 with the proposal of Pauli in order to explain the continuous β-spectrum. He proposed that in the β-decay process, the electron is emitted together with a massless and chargeless particle of spin 1/2. He gave the name neutron (the neutral one) to this particle. The particle that today we call neutron, was discovered in 1932 by J. Chadwick. After the discovery of neutron, the particle proposed by Pauli was called neutrino (little neutron) by Enrico Fermi.

Fermi, in 1934, gave the first successful theory of β- decay. Experimental detection of neutrino was done by F. Reines and C. W. Cowan at Savanah River experiment during 1954-1956. In 1962,

Leon M. Lederman, Melvin Schwartz and Jack Steinberger showed that more than one type of neutrino exists by first detecting interactions of the muon neutrino. When a third type lepton the tau (τ), was discovered in 1975 at the Stanford Linear Accelerator, it too was expected to have an associated neutrino. First evidence for this third neutrino type came from the observation of missing energy and momentum in β decays analogous to the β-decay that had led to the discovery o[the neutrino in the first place. The first detection of actual tau neutrino (n_τ) interactions was announced in summer of 2000 by the DONUT collaboration at Fermi lab. Since neutrinos arc extremely difficult to detect because they hardly interact with matter at all - only one in a million of the neutrinos traversing earth collide with an atomic nucleus, hence, after decades of painstaking experimental and theoretical work, neutrinos have now become enshrined as an essential part of the accepted quantum description of fundamental particles and forces, the Standard Model of Particle Physics. This is a highly successful theory in which elementary building blocks of matter are divided into three generations of two kinds of particle - quarks and leptons, which may be summarized as:

Generation	**I**	**II**	**III**
Quarks	$\binom{u}{d}$	$\binom{c}{s}$	$\binom{t}{b}$
Lepton	$\binom{\nu_e}{e^-}$	$\binom{\nu_\mu}{\mu^-}$	$\binom{\nu_\tau}{\tau^-}$

Netutrino Oscillations

The first suggestion that free neutrinos traveling through space might oscillate i.e. periodically change from one neutrino type to another was made in 1957 by Bruno Pontecorvo who proposed $\nu \rightarrow \overline{\nu}$ oscillations in analogy with $K \rightarrow \overline{k}$ oscillations, described as the mixing of two Majorana neutrinos. Majorana neutrinos are defined as the particles which are equivalent to their own antiparticles. Pontecorvo was the first to realize that what we call the "electron neutrino", for example, may be a linear combination of mass eigenstate neutrinos and that this feature could lead to neutrino oscillations of the kind $\nu_e \nu_\mu$.

Solar Neutrinos

The first clues that neutrinos have mass came from an experiment deep underground, carried out by an American scientist Raymond Davis Jr., detecting solar neutrinos. It revealed only about one-third of the number predicted by theories of how the sun works pioneered by John Bahcall. The result puzzled both solar and neutrino physicists. Some Russian researchers, Mikheyev and Smirnov, developing ideas proposed previously by Wolfenstein in the U.S., suggested that the solar neutrinos might be changing into something else. Only electron neutrinos are emitted by the sun through the reaction

$$p + p \rightarrow D + e^+ + \nu_e \qquad (1)$$

and they could be converting into muon and tau neutrinos which were not being detected by the experiments.

The precise mechanism for solar neutrino oscillation proposed by Mikheyev, Smirnov and Wolfenstein involved the resonant enhancement of neutrino oscillations due to matter effects. Just as light passing through matter slows down, which is equivalent to the photon gaining a small effective mass, neutrinos passing through matter also result in the neutrinos slowing down and gaining a small effective mass. The effective neutrino mass is largest when the matter density is highest which in the case of solar neutrinos is in the core of the sun. In particular electron neutrinos generated in the core of the sun will be subject to such matter effects. It turns out that neutrino oscillations, which would be present in the vacuum due to neutrino mass and mixing, will exhibit strong resonant effects in the presence of matter as the effective mass of the neutrinos varies along the path length of the neutrinos. This can result in a resonant enhancement of solar neutrino oscillations known as the MSW effect.

Atmospheric Neutrinos

In 1992, another neutrino deficit was observed in the ratio of muon neutrinos to electrons neutrinos produced at the top of the earth's atmosphere. When high-energy cosmic rays mostly protons, strike nuclei in the upper atmosphere, they produce pions and muons which then decay through the weak force and produce muon and electron neutrinos. This is called the atmospheric neutrino oscillation.

$$\pi^+ \rightarrow \mu^+ + \mathrm{v}_\mu$$

$$\mu^+ \rightarrow \mathrm{e}^+ + \mathrm{v}_\mathrm{e} + \overline{\mathrm{v}}_\mu \tag{2}$$

and

$$\pi \rightarrow \mu + \overline{\mathrm{v}}_\mu$$

$$\mu \rightarrow \mathrm{e} + \overline{\mathrm{v}}_\mathrm{e} + \mathrm{v}_\mu \tag{3}$$

The atmospheric neutrinos have very high energies, ranging from hundreds of mega-electron volt (MeV) to tens of giga-electrons-volt (GeV), depending on the fragments of initial reaction.

Reactor Neutrinos

Reactor experiments are disappearance experiments looking for $(\downarrow \mathrm{e} \rightarrow (\overline{\downarrow}\, \mathrm{x}$. In reactor experiments one tries to prove that lesser number of neutrinos of the same flavor reach the detector than what would be expected from the source. The disadvantage of disappearance experiments is that it is necessary to have a precise knowledge of the neutrino flux. The advantage is that they measure all channels, e.g. $(\downarrow^\mathrm{e} \rightarrow (\downarrow \mu, (\downarrow \tau$ etc. In this class, the most promising ones are investigations of solar neutrinos, atmospheric neutrinos, neutrinos from supernova explosions and the neutrino cosmic background radiation yet to be observed.

Several reactor experiments have been done in the past namely ILL-Grenoble, Bugey, Rovno, Savannah River, Gosgen, Krasnojarsk and Buge III. Two new reactor experiments are

CHOOZ and Palo Verdi. The CHOOZ experiment (France) is located underground with a shielding of 300 m.w.e. reducing the background due to atmospheric muons by a factor of 300. Moreover, the detector is about 1030 m away from two 4.2 GW reactors enlarging the sensitivity to smaller Äm². The Palo Verdi experiment near Phocnix, AZ (USA) consists of 12 ton liquid scintillator also loaded with Gd. The experiment is located under a shielding of 46 m.w.e. in a distance of about 750 (829) m to three reactors with a total power of 10.2 GW.

Mechanics of Neutrino Oscillations

Oscillation, or the spontaneous periodic change of one neutrino mass state to another, can beautifully be explained by quantum mechanics. A neutrino produced through the weak force (for example in muon decay) is described as the sum of two matter waves. As the neutrino travels through space, these matter waves interfere with each other constructively or destructively. This interference causes first the disappearance and then the reappearance of the original type of neutrino. The interference can occur only if the two matter waves have different masses. Thus the mechanics of oscillation starts from the assumption that the lepton weak and mass states are not same and one set is composed of mixture of the other set.

Let we have a source of v_l neutrinos. The flavor eigenstate v_l is not a mass eigen state but a linear superposition of mass v_m. Assuming that there are N number of mass eigenstates eigenstates v_m, each neutrino produced by the source is coherent superposition of the mass states

$$v_l = \sum_{m=1}^{N} U_{lm} v_m \tag{4}$$

Here U_{lm} is the neutrino mixing matrix, defined later.

If a neutrino v_l borns at time t = 0 with momentum P_v, then at this time the wave function is given by

$$\psi(x, t=0) = \sum_m U_{lm} v_m e^{ipvx} \tag{5}$$

After a time t this will evolve into,

$$\psi(x,t) = \sum_{m} U_{lm} v_m e^{ipvx} e^{-iEmt} \quad (6)$$

With $$E_m = [E(v]_m) = \sqrt{p_v^2 + M_m^2} \quad (7)$$

Where M_m is the mass of neutrino mass eigenstate V_m.

$$E_m = p_v + \frac{M_m^2}{2p_v} \quad (8)$$

With all $M_m << P_v$, neutrino travels almost with the speed of light. Hence, if it was born at x = 0 then after time t, it will be approximately at x = t so that ψ(t, t) = ψ(x, x) which is given by

$$\psi(x,x) \approx \sum_{m} U_{lm} v_m e^{-i\left(M_m^2 / 2pv\right)x} \quad (9)$$

Expressing V_m as a combination v_ls we find

$$v_m = \sum_{\ell} U^{*'}\ell m v_\ell ' \quad (10)$$

so that

$$\psi(x,x) \square \sum_{\ell} \left[\sum_{m} U_{\ell m} e^{-i\left(M_m^2 / 2pv\right)x} U'_\ell m^* \right] v'l \quad (11)$$

This wave function is a superposition of all the neutrino flavors. In it the amplitude for our neutrino born with flavor l to have a new flavor l' after travelling a distance x *is just the coefficient of* $v'_{l'}$

Let $p(l \to l', x)$ be the probability of finding the neutrino to have a flavor l, at a distance x from its source, if originally it had flavor *l*, then from Eqn. (11)

$$p(\ell \to \ell', x) = \left[\sum_{m} U'^*_{lm} e^{i\,''} \right] \cdot \left[\sum_{m} U_{\ell m} e^{-1} \left(M_m^2 / 2pm \right) x U'_l m^* \right] \quad (12)$$

$$= \sum [|U_{lm}|^2 |U'_{\ell m}|^2 + \sum_{m'=m} \mathrm{Re}\left(U_{\ell m} U^{*\prime}{}_{\ell m}\, U'_{\ell}\, m' U^{*\prime}{}_{\ell} m\right) \cos\left(\frac{M'^2_m}{2p_v} x\right)$$

$$+ \sum_{m'=m} \mathrm{Im}(U_{\ell m} U^{*\prime}{}_{lm} U'_{\ell}\, m' U^{*\prime}{}_{\ell} m) \sin\left(\frac{M^2_m - M'^2_m}{2p_v} x\right) \quad (13)$$

Now if CP is conserved, then U is real and above equation simplifies to

$$p(\ell \to \ell', x) = \sum_m U^2_{\ell m} U'_{\ell} m^2 + \sum_{m'=m} (U_{\ell m} U'_{\ell m} U'_{\ell} m' U'_{\ell} m) \cos\left(2\pi \frac{x}{L'_{mm}}\right) \quad (14)$$

This probability has an oscillatory pattern as a function of distance x.

In this equation, the quantity L_{mm} is the oscillation length between V_m and V_m' and is given by

$$L_{mm'} = 2\Pi \frac{2p_v}{|M^2_m - M'^2_m|} \equiv 2\Pi \frac{2p_v}{|\delta M'^2_{mm}|} \quad (15)$$

It is noted that-

- If all the masses are equal (in particular, if they all vanish), then there is no oscillation.
- If $v_{\ell o} = v_{mo,} v_{\ell o}$ doesn't oscillate into v_{ℓ} with $\ell' \neq \ell_{o.}$

That is, oscillation requires neutrinos to have both mass and non-trivial mixing.

Neutrino Masses and Mixing Angles

Three State Neutrino Mixing

The minimal neutrino sector required to account for the atmospheric and solar neutrino oscillation data consists of three light physical neutrinos with left-handed flavor eigenstates, ν_e, ν_μ and ν_τ defined to be those states that share the same doublet as the charged lepton mass eigenstates e, μ and τ. Within the framework of three–neutrino oscillations, the neutrino flavor eigenstates ν_e, ν_μ and ν_τ are related to the neutrino mass eigenstates ν_1, ν_2 and ν_3 with masses m_1, m_2 and m_3 respectively by a 3 × 3 unitary matrix called the lepton mixing matrix U_{ei}.

$$\begin{pmatrix} v_e \\ v_\mu \\ v_\tau \end{pmatrix} = \begin{pmatrix} U_{e1} & U_{e2} & U_{e3} \\ U_{\mu 1} & U_{\mu 2} & U_{\mu 3} \\ U_{\tau 1} & U_{\tau 2} & U_{\tau 3} \end{pmatrix} \begin{pmatrix} v_1 \\ v_2 \\ v_3 \end{pmatrix} \qquad (16)$$

If the light neutrinos are Majorana, U_{ei} (i=1,2 & 3) can be parameterized in terms of three mixing angles θ_{ij} and and three complex phases. A unitary matrix has six phases but three of them are removed by the phase symmetry of the charged lepton Dirac masses. Since the neutrino masses are Majorana, there is no additional phase symmetry associated with them unlike the case of quark mixing where a further two phases may be removed. If we begin by assuming that the phases are zero, the lepton mixing matrix U_{ei} may be parameterized by a product of three Euler rotations as depicted in the following Fig. 1.

The mixing matrix U_{ei} is further given by a product of three matrices-

$$U_{ei} = R_{23} R_{13} R_{12} \qquad (17)$$

Where

$$R_{23} = \begin{pmatrix} 1 & 0 & 0 \\ 0 & c_{23} & s_{23} \\ 0 & s_{23} & c_{23} \end{pmatrix}, R_{13} = \begin{pmatrix} c_{13} & 0 & s_{13} \\ 0 & 1 & 0 \\ s_{13} & 0 & c_{13} \end{pmatrix}, R_{12} = \begin{pmatrix} c_{12} & s_{12} & 0 \\ s_{12} & c_{12} & 0 \\ 0 & 0 & 1 \end{pmatrix}$$

On simplification we get

$$U_{ei} = \begin{pmatrix} c_{13}c_{12} & c_{13}s_{12} & s_{13} \\ -c_{12}s_{23}s_{13} - c_{23}s_{12} & -s_{12}s_{23}s_{13} + c_{23}c_{12} & s_{23}c_{13} \\ -c_{23}s_{13}c_{12} + s_{23}s_{12} & -c_{23}s_{13}s_{12} - s_{23}c_{12} & c_{23}c_{13} \end{pmatrix} \qquad (18)$$

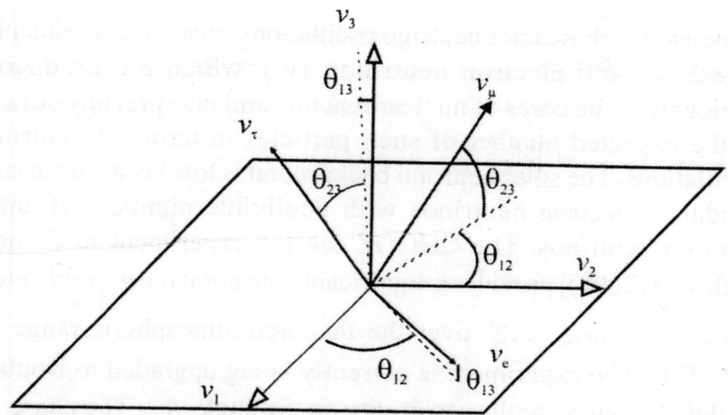

Fig. 1: The Relation Between the Neutrino Weak Eigenstates v_e, v_m and v_t and the Mass Eigenstates v_1, v_2 and v_3 in Terms of the Three Mixing Angles q_{12}, q_{13} and q_{23}.

In the above matrix, $C_{ij} = \cos\theta_{ij}$ and $s_{ij} = \sin\theta_{ij}$. The allowed range of the angles is $0 \leq \theta ij \geq \pi/2$. Ignoring phases, these angles are just the Euler angles representing the rotation of one orthogonal basis into another. Including phases, the lepton mixing matrix is summarized in Fig. 2 shown below.

$$U_{MNS} = \begin{pmatrix} 1 & 0 & 0 \\ 0 & c_{23} & s_{23} \\ 0 & -s_{23} & c_{23} \end{pmatrix} \begin{pmatrix} c_{13} & 0 & s_{13}e_{18} \\ 0 & 1 & 0 \\ -s_{13}e^{18} & 0 & c_{13} \end{pmatrix} \begin{pmatrix} c_{12} & c_{12} & 0 \\ -s_{12} & c_{12} & 1 \\ 0 & 0 & 1 \end{pmatrix} \begin{pmatrix} e^{la\frac{1}{2}} & 0 & 0 \\ 0 & e^{la\frac{2}{2}} & 0 \\ 0 & 0 & 1 \end{pmatrix}$$

Atmospheric | Reactor | Solar | Majorana

Fig. 2: The Lepton Mixing Matrix, Factorized into Matrix Product of Four Matrices

The phases $\alpha_{1,2}$ are called Majorana phases since they are only present if the neutrino mass is Majorana. The phase δ is called the Dirac phase since it is always present even if neutrinos have Dirac mass.

The first matrix in Fig. 2 is associated with atmospheric neutrino oscillations. The physics of the second matrix in Fig. 2 is associated with Reactor neutrino oscillations. Reactor experiments detect the anti-electron neutrinos ($\bar{\nu}_e$) which are produced copiously in the cores of nuclear reactors and interpret any deficit in the expected number of such particles in terms of neutrino oscillations. The solar neutrino background is low because the sun produces electron neutrinos with negligible numbers of anti-electron neutrinos. The CHOOZ reactor experiment in France [Schwetz (2008)] provides a significant constraint on θ_{13} *and limits* $sin\,\theta_{12} \leq 0.23$ or $\theta_{13} \leq 12^*$ over the favoured atmospheric range at 90% C.L. The experiment is currently being upgraded to Double CHOOZ to increase the sensitivity on the angle θ_{13}. The phase δ also appears in the third matrix and physically represents CP violation. The angle θ_{13} has not yet been measured but a huge experimental effort is under way to measure both the angle θ_{13} and the CP phase δ. However it should be emphasized that the CP-violation in the lepton sector is one of the most challenging frontiers in the future studies of neutrino mixing.

The physics of the third matrix in Fig. 2 is associated with solar neutrino oscillations as discussed above and recently confirmed by the Japanese reactor experiment KamLAND that measures $\bar{\nu}_e$'s produced by several surrounding nuclear reactors.

The physics of the fourth matrix in Fig. 2 is associated with Majorana neutrino masses. These phases could in principle be measured in neutrinoless double beta ($0\nu\ \beta\beta$) decay experiments.

Experimental Aspects

There are various neutrino oscillation experiments that are running at present. Basically, they can be categorized into following two types, looking for neutrino oscillations:

- ***Disappearance Experiments*:** Here, one looks for a reduction in the expected number of the neutrinos to be detected of a particular flavour. Then this disappearance is explained in terms of neutrino oscillations into undetected flavours.

- ***Appearance Experiments***: Here, one looks for neutrino flavours which are either not present or very weakly produced at the neutrino source. Again this appearance is explained in terms of neutrino oscillations.

Below, we are presenting a brief discussion on present ongoing and future planned neutrino oscillation experiments.

Ongoing Experiments

Some of these experiments are Super-Kamiokande, SNO, LSND, MiniBooNE, KamLAND, CHOOZ etc. which measure the fluxes of solar and atmospheric neutrinos and thereby put limits on corresponding mixing angles.

SK (Super-Kamiokande)

Super-Kamiokande is a disappearance type of Japanese experiment which measures the solar and atmospheric neutrinos. The Super-Kamiokande experiment consists of thousands of tons of pure water in a tank deep underground and was originally built to search for proton decay but because of its design, it was also able to detect highly energetic neutrinos from the sun that interact with electrons via scattering reactions. These electrons can travel faster than the local speed of light in water causing them to emit the optical equivalent of a sonic boom – a glow of blue light called Cerenkov radiation that can be detected by ultra-sensitive photomultiplier tubes around the tank. Super-Kamiokande also measured the number of electron and muon neutrinos that arrive at the Earth's surface as a result of cosmic ray interactions in the upper atmosphere, which are referred to as 'atmospheric neutrinos'. While the number and angular distribution of electron neutrinos is as expected, Super-Kamiokande showed that the number of muon neutrinos is significantly smaller than expected and that the flux of muon neutrinos exhibits a strong dependence on the zenith angle. These observations gave compelling evidence that muon neutrinos undergo flavour oscillations and this in turn implies that at least one neutrino flavour has a non-zero mass. The standard interpretation, well supported by current data is that muon neutrinos are oscillating into tau neutrinos.

SNO (Sudbury Neutrino Observatory)

Sudbury Neutrino Observatory (SNO) in Canada is a water Cerenkov detector like Super-Kamiokande, but instead of using normal water it uses heavy water D_2O. The deuterons (D) in the heavy water are the most weakly bound of all nuclei which gives SNO the chance to observe three different reactions induced by solar neutrinos. The first of these processes is the charged-current (CC) reaction, $v_e + D \rightarrow p + p + e^-$, which is detected by observing Cerenkov photons from the energetic recoil electron e^-. Second is the neutral-current (NC) reaction, $\nu_\alpha + D \rightarrow p + n + \nu_\alpha$. This is observed via the emitted neutrons n and is independent of the flavor of the incoming neutrino n_a. It therefore provides a way to normalize the total flux of neutrinos being emitted by the sun. Finally SNO measures the process of elastic scattering (ES) reaction, also measured in Super-Kamiokande, $\nu_\alpha + e^- \rightarrow \nu_\alpha + e^-$, which has some sensitivity to all neutrino flavours. SNO measurements of CC reaction on deuterium are sensitive exclusively to v_e's, while the ES off electrons has a small sensitivity to ν_μ's and ν_τ's. The CC ratio is significantly smaller than the ES ratio. This immediately disfavours oscillations of v_e's to sterile neutrinos which would lead to a diminished flux of electron neutrinos but equal CC and ES ratios. On the other hand, the different ratios are consistent with oscillations of v_e's to active neutrinos ν_μ's and ν_τ's since this would lead to a larger ES rate as this has a neutral current component. The SNO analysis is nicely consistent with both the hypothesis that electron neutrinos from the sun oscillate into other active flavors and with the Standard Solar Model prediction. The latest results from SNO including the data taken with salt inserted into the detector to boost the efficiency of detecting the neutral current events strongly favour the large solar mixing angle (LMA) MSW solution.

LSND (Liquid Scintillator Neutrino Detector)

The LSND based at Los Alamos National Laboratory (LANL) in the United States, was an appearance type accelerator experiment. LSND had a base-line of 30 m and looked for an excess of ν_e, $\overline{\nu}_e$ starting from a beam which is mainly made up of ν_μ (and $\overline{\nu}_\mu$). In LSND, the source of neutrinos was an intense proton beam at the Los Alamos Meson Physics Facility (LAMPF) whose kinetic energy was 800 MeV (~1 mA current). This beam was made to hit a water target, followed by a water-cooled copper (Cu) beam dump. This produces a large number of pions, mostly π^+. The π^+ decays into μ^+ and ν_μ as $\mu^+ \rightarrow e^+ + \nu_e + \overline{\nu}_\mu$. The detector was a cylindrical tank containing 167 t of mineral oil (CH2) in an organic scintillating medium and was lined up with 1220 phototubes inside the tank to detect the Cherenkov radiation as well as the scintillation light emitted from the propagating particles inside it.

LSND collected data from 1993 up to 1998 and the collaboration found an excess of electron antineutrinos from a beam of neutrinos consisting of the decay products of pion particle beam. The conclusion was that muon antineutrinos in the beam were changing into electron antineutrinos while propagating. This would either require a further light neutrino state with no weak interactions called 'sterile neutrino' or some other non-standard physics. The collaboration first reported evidence for anti-neutrino oscillations in 1995, thus becoming the first experiment to report observation of neutrino oscillations using appearance type strategy.

MiniBooNE

In order to address the LSND anomaly (regarding the accuracy about background estimates, etc.) the MiniBooNE (BooNE stands for the Booster Neutrino Experiment) experiment was proposed. The MiniBooNE collaborators kept the L/E the same as in LSND but changed the systematic, energy and event signature. MiniBooNE is located at the Fermi National Accelerator Laboratory in the United States. The experiment made use of the Fermilab Booster neutrino beam. Protons with energies of 8 GeV were incident on a beryllium target. This experiment was also based on the 'appearance' Principle. It had looked for an excess of ν_e in a purely ν_μ beam. After the protons hit the target, the produced (positively charged) pions and kaons pass through a collimator of about 60 cm long and then through a tunnel towards the detector which is about 50 m long. These particles decay along the way producing neutrinos. The produced neutrinos traverse along the tunnel, enter the detector, and interact with the medium in the detector. Depending on the pattern of light observed in the PMTs, one can determine the kind of interaction the neutrino went through in the detector. Cherenkov radiation and scintillation (fluorescence) light were used to detect the kind of neutrino interaction. Neutrinos interact through both charged current and neutral current channels and both were used in the detection process. Data were collected for about five years starting from 2002. The MiniBooNE results rule out the simplest sterile neutrino schemes with single sterile neutrino as these are not compatible with both LSND and MiniBooNE data. Moreover, MiniBooNE excludes the simplest two neutrino oscillation interpretation of the LSND signal at 98% C.L.

KamLAND

KamLAND is a Japanese reactor experiment that measures electron antineutrinos produced by several surrounding nuclear reactors. KamLAND consists of 1 kton of ultra-pure liquid scintillator (LS) contained in a 13-m-diameter transparent nylonbased balloon suspended in non-scintillating oil. The balloon is surrounded by 1879 photomultiplier tubes (PMTs) mounted on the inner surface of an 18-m-diameter spherical stainless steel vessel. Electron anti-neutrinos are detected via inverse β decay, $\overline{\nu}_e + p \rightarrow e^+ + n$, with a 1.8 MeV $\overline{\nu}_e$ energy threshold. KamLAND has already seen a signal of neutrino oscillation over the solar neutrino LMA MSW mass range and has recently confirmed the LMA MSW region.

Apart of these experiments, neutrino oscillations consistent with atmospheric neutrino observations have been seen using neutrino beams fired over hundreds of kilometers as in the K2K experiment in Japan, the Fermilab-MINOS experiment in the US or the CERN-OPERA experiment in Europe. Further long-baseline neutrino beam experiments are in the pipeline and neutrino oscillation physics is poised to enter the precision era with Super beams and a Neutrino Factory on the horizon.

Planned Experiments

It is expected that future neutrino oscillation experiments will give accurate information about the mass squared splittings, $\Delta^{m_{ij}^2} = m_i^2 - m_j^2$, mixing angles and the CP violating phase δ. Long baseline neutrino beam experiments will give more accurate determinations of the atmospheric parameters, eventually to 10%.

One of the oscillation experiments called MINOS (Main Injector Neutrino Oscillation Search) was proposed in 1995 with a neutrino beam pointed from Fermilab to the Soudan mine in Minnesota with a baseline of 735 km. The experiment started running in the spring of 2005 and within a year had gathered data corresponding to 1.27×10^{20} protons on target. MINOS will run for 5 years with a goal of accumulating 16×10^{20} protons on target which should improve our knowledge of the oscillation parameters. In addition, a neutrino beam from CERN to the OPERA detector in the Gran Sasso tunnel is presently underway and experimenters are looking for tau neutrinos (τ) tracks to prove conclusively that muon neutrinos oscillate to tau neutrinos.

In the next couple of years T2K, a Japanese experiment sending a neutrino beam from the J-PARC complex to Super-Kamiokande is due to start. It will be an 'off-axis superbeam' over a baseline of 295 km. Neutrino beams originate from charged pion decays which generally results in a large spread of neutrino energies. However, for a specific angle relative to the pion direction, the neutrinos have a quite monochromatic energy spectrum and therefore such 'off-axis' neutrino beams will have quite a well defined energy which can be advantageous for certain experimental measurements. Its first goal is to measure θ_{13} or set a limit on it of about 0.05 (as compared to the CHOOZ limit on θ_{13} of about 0.2). Interestingly MINOS over a LBL of 735 km is more sensitive than J-PARC to matter effects, so there should be some interesting complementarity between these two experiments which could for example allow the sign of Δm_{32}^2 to be determined. An off-axis superbeam version of the MINOS experiment called NOνA is seeking approval in the US.

The ultimate goal of oscillation experiments however is to measure the CP violating phase δ. To do this it would seem that all the stops would need to be pulled out in neutrino physics experiments. Various Superbeam, or Beta-beam or Neutrino Factory options are currently being considered. For example an upgraded J-PARC with a 4MW proton driver and a 1 megaton Hyper-Kamiokande detector or some sort of Neutrino ν_μ Factory based on muon storage rings would seem to be required for this purpose.

Neutrino Mass Hierarchy

From neutrino oscillation data, it follows that solar mass squared difference is much smaller than the atmospheric one. For three massive neutrinos, two types of neutrino mass spectra are possible:

1. Normal mass spectrum:

$$m_1 < m_2 < m_3; \qquad \Delta m_{21}^2 \ll \Delta m_{31}^2$$

Where $\quad \Delta m_{21}^2 = m_2^2 - m_1^2 \quad$ and $\quad \Delta m_{31}^2 = m_3^2 - m_1^2$

2. Inverted mass spectrum:

$$m_3 < m_1 < m_2; \qquad \Delta m_{21}^2 \ll \Delta m_{13}^2$$

here $\quad \Delta m_{21}^2 = m_2^2 - m_1^2 \quad$ and $\quad \Delta m_{13}^2 = m_1^2 - m_3^2$

In the case of normal spectrum the neutrino masses $m_{2,3}$ are connected with the lightest mass m_1 and the two neutrino mass squared differences Δm_{21}^2 and Δm_{31}^2 by the following relations

$$m_2 = \sqrt{m_1^2 + \Delta m_{21}^2}, \quad m_2 = \sqrt{m_1^2 + \Delta m_{31}^2}$$

In case of inverted spectrum, lightest mass state m_3 is related to other two mass states m_1 and m_2 by the following relation

$$m_1 = \sqrt{m_3^2 + \Delta m_{13}^2}, \qquad m_2 = \sqrt{m_3^2 + \Delta m_{13}^2 + \Delta m_{21}^2},$$

It is seen that effective Majorana mass is determined not only by the lightest neutrino mass and neutrino mass squared differences but also by the character of the neutrino mass spectrum.

Usually the following three typical neutrino mass spectra are considered:

1. Normal hierarchy of neutrino masses: $m_1 < m_2 < m_3$
2. Inverted hierarchy of neutrino masses: $m_3 \leq m_1 < m_2$
3. Quasi-degenerate neutrino mass spectrum: $m_1 \approx m_2 \approx m_3$

Evaluation of Effective Neutrino MASS$\langle m_v \rangle$

Following the notation of Avignone et al., the effective mass of neutrino $<M_v>$ is given by elements of I[st] row of neutrino mixing matrix U_{ei} defined above,

$$|< m_v >| \equiv \| [U_{e1}^L| \,]^2 m_1 + |U_{e2}^L|]^2 e^{i\ell z} m_2 + |U_{e3}^L|]^2 e^{i\ell 3} m_3 |$$

$$= |c_3^2 c_2^2 m_1 + c_3^2 s_2^2 e^{i\upsilon 2} m_2 + S_3^2 e^{i\upsilon 3} m_3| \quad (19)$$

Where $e^{i\upsilon}2, e^{i\upsilon}3$ are Majorana phases.

Let

$$A= \left|[U_\downarrow e1^\uparrow L|\right]^\uparrow 2c_\downarrow 2^\uparrow 2,\ B=\left|[U_\downarrow e2^\uparrow L|\right]^\uparrow 2=c_\downarrow 3^\uparrow 2s_\downarrow 2^\uparrow 2,\ c=\left|[U_\downarrow e3^\uparrow L|\right]^{\uparrow 2} =s_\downarrow 3^\uparrow 2$$

Then after some calculation, $<m_v>^2$ comes out to be

$$|< m_v >|^2 = A^2 m_1^2 + B^2 m_2^2 + C^2 m_3^2 + 2ABm_1 m(\cos\varphi_3) +$$

$$2BCM_2 m_3 \cos[(\varphi]_2 - \varphi_3) \quad (20)$$

From Eqn. (20), $<m_v>$ can be solved for the following three types of neutrino mass hierarchy-

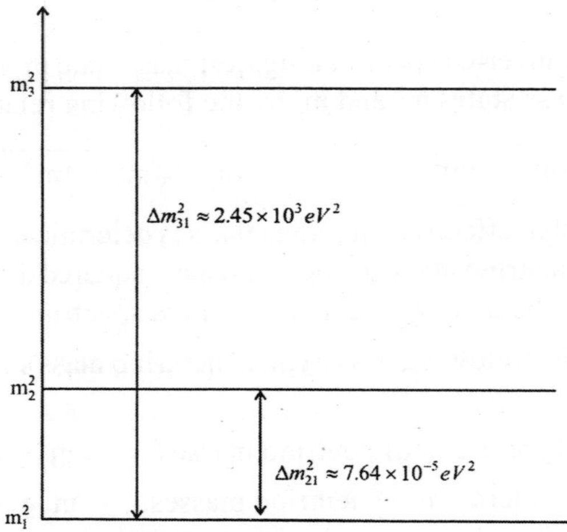

Fig. 3: Normal Hierarchy Mass Spectrum

1. Normal mass hierarchy

   This corresponds to the case

$$m_1 \ll m_2 < m_3 \Delta m_s^2 = \Delta m_{21}^2 = m_2^2 - m_1^2 \Rightarrow m_2 = \sqrt{m_1^2 + \Delta m_s^2} \quad (21)$$

$$\Delta m_A^2 = \Delta m_{31}^2 = m_3^2 - m_1^2 \Rightarrow m_3 = \sqrt{m_1^2 + \Delta m_A^2} \quad (22)$$

Δm_s^2 and Δm_A^2 are experimentally determined solar and atmospheric neutrino parameters.

Substituting the values of m_2 and m_3 in Eqn. (20), we get

$$< m_v >^2 = A^2 m_1^2 + B^2(\Delta m_s^2 + m_1^2) + C^2(\Delta m_A^2 + m_1^2) + 2ABm_1$$

$$(\Delta m_s^2 + m_1^2)^{1/2}(\cos\varphi_2) + 2ACm_1(\Delta m_A^2 + m_1^2)^{1/2}(\cos\varphi_3) + 2BC$$

$$(\Delta m_s^2 + m_1^2)^{1/2}(\Delta m_A^2 + m_1^2)^{1/2}\cos[\cos(\varphi]_2 - \varphi_{3)} \quad (23)$$

2. Inverted mass hierarchy

   This corresponds to the case

$$-\Delta m_A^2 = \Delta m_{31}^2 = m_3^2 - m_1^2 \Rightarrow m_1 = \sqrt{m_3^2 + \Delta m_A^2} \quad (24)$$

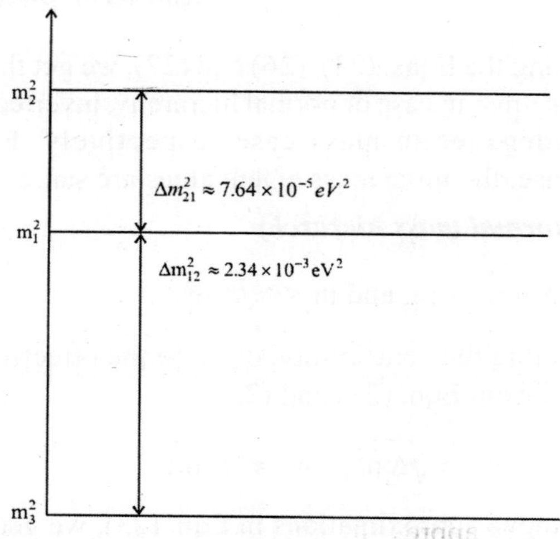

Fig. 4: Inverted Hierarchy Mass Spectrum

$$\Delta m_s^2 = \Delta m_{21}^2 = m_2^2 - m_1^2 \quad \Rightarrow m_2 = \sqrt{m_3^2 + \Delta m_A^2 + \Delta m_s^2} \quad (25)$$

Substituting the values of m_1 and m_2, Eqn (20) simplifies to

$$\langle m_\nu \rangle^2 = A^2(m_3^2 + \Delta m_A^2) + B^2(m_3^2 + \Delta m_A^2 + \Delta m_s^2) + c + c^2 m_3^2 + 2AB(m_3^2 + \Delta m_A^2)^{\frac{1}{2}}(m_3^2 + \Delta m_A^2 + \Delta m_s^2)^{1/2}(\cos\varphi_2) + 2AC(m_3^2 + \Delta m_A^2)^{\frac{1}{2}} m_3(\cos\varphi_3) + 2BC(m_3^2 + \Delta m_A^2 + \Delta m_s^2)^{\frac{1}{2}} m_3 \cos[(\varphi]_2 - \varphi_3) \quad (26)$$

3. Almost degenerate (AD) mass hierarchy

For this type of mass hierarchy, we have

$$m_1 = m_2 = m_3$$

Hence from Eqn. (20), we have

$$< m_\nu >^2 = A^2 m_1^2 + B^2 m_1^2 + C^2 m_1^2 + 2ABm_1^2 \cos\varphi_2 + 2ACm_1^2 \cos\varphi_3 + 2BCm_1^2 \cos[(\varphi]_2 - \varphi_3) = m_\downarrow 1^\uparrow 2 + B^\uparrow 2 + C^\uparrow 2 + 2AB\cos\varphi_\downarrow 2 + 2CA\cos\varphi_\downarrow 3 + 2BC\cos(\varphi_2 - \varphi_3)] \quad (27)$$

On solving the Eqns. (23), (26) and (27), we get the effective neutrino mass $<m_i>$ in case of normal hierarchy, inverted hierarchy and almost degenerate mass case respectively. For almost degenerate case, the three mass eigenvalues are same.

$<m_v>$ from normal mass hierarchy

In this case $m_1 \ll m_2 < m_3$ and $m_1 << \sqrt{\Delta m_s^2}$

so that neglecting the contribution of m_1 to the effective neutrino mass, we find from Eqn. (21) and (22),

$$m_2 \approx \sqrt{\Delta m_S^2}, \quad m_3 \approx \sqrt{\Delta m_A^2} \tag{28}$$

Using above approximations in Eqn. (23), we get following upper and lower bounds for effective majorana neutrino mass

$$< m_v > \leq \left(\cos^2\theta_{13}\sin^2\theta_{12}\sqrt{\Delta m_S^2} + \sin^2\theta_{13}\sqrt{\Delta m_A^2}\right)$$

or

$$\langle m_{\downarrow} v\rangle_{\downarrow} \max^{\uparrow} NH \approx (\cos^{\uparrow} 2\theta_{\downarrow} 13\sin^{\uparrow} 2\theta_{\downarrow} 12\sqrt{}c[\Delta m]_{\downarrow} s^{\uparrow} 2) + \sin^{\uparrow} 2\theta_{\downarrow} 13\sqrt{}([\Delta m]_{\downarrow} A^{\uparrow} 2)) \tag{29}$$

and

$$\langle m_v\rangle \geq \left|\cos^2\theta_{13} som^2\theta_{12}\sqrt{\Delta m_S^2} - \sin^2\theta_{13}\sqrt{\Delta m_A^2}\,\right|$$

or

$$[\langle m_{\downarrow} v\rangle]_{\downarrow} \min^{\uparrow} NH \approx (\cos^{\uparrow} 2\theta_{\downarrow} 13\sin^{\uparrow} 2\theta_{\downarrow} 12\sqrt{}([\Delta m]_{\downarrow} s^{\uparrow} 2) - \sin^{\uparrow} 2\theta_{\downarrow} 13\sqrt{}([\Delta m]_{\downarrow} A^{\uparrow} 2)) \tag{29}$$

and

$$\langle m_v\rangle\left|\cos^2\theta_{13}\sin^2\theta_{12}\sqrt{\Delta m_s^2} - \sin^2\theta_{13}\sqrt{\Delta m_A^2}\right|$$

or

$$[\langle m_{\downarrow} v\rangle]_{\downarrow} \min^{\uparrow} NH \approx$$

$$\left|\cos^{\uparrow} 2\theta_{\downarrow} 13\sin^{\uparrow} 2\theta_{\downarrow} 12\sqrt{}([\Delta m]_{\downarrow} s^{\uparrow} 2) - \sin^{\uparrow} 2\theta_{\downarrow} 13\sqrt{}([\Delta m]_{\downarrow} A^{\uparrow} 2)\right. \tag{30}$$

We have used, $e^{i\varphi 2} = e^{i\varphi 3} = +1$ for upper bound and $e^{i\varphi 2} = -e^{i\varphi 3} = +1$ for lower bound in Eqn. (23)

Using the best fit values of solar and atmospheric neutrino parameters and SBL angle θ_{13} *[Schwetz* et al. *(2011)]*-

$$\Delta m_s^2 = 7.64 \times 10^{-5} EV^2, \qquad \Delta m_A^2 = 2.45 \times 10^{-3} EV^2$$

$$\sin^2\theta_{12} = \sin^2\theta_{sol} = 0.316, \qquad \sin^2\theta_{12} = 0.017 \tag{31}$$

we find, $\cos^2\theta_{13}\sin^2\theta_{12}\sqrt{\Delta m_s^2} \approx 2.715 \times 10^{-3} eV$

and $\sin^2\theta_{13}\sqrt{\Delta m_A^2} \leq 0.842 \times 10^{-3} eV$

so that $\langle m_\downarrow v\rangle_\downarrow \max^\uparrow NH \approx 3.56 \times 10^{-3} eV$

and $[\langle m_\downarrow v\rangle]_\downarrow \max^\uparrow NH \approx 1.67 \times 10^{-3} eV$ (32)

For the choice of three possible values of $\sin^2\theta_{13} = 0.009$, 0.002 and 0.00, comparable to CHOOZ upper bound (< 0.009), the following intervals are obtained for effective mass $\langle m_\nu\rangle$:

$$\sin^2\theta_{13} = 0.0009 \Rightarrow \quad 2.09 \times 10^{-3} eV \leq \langle m_\nu\rangle \leq 3.18 \times 10^{-3} eV$$

$$= 0.002 \Rightarrow \quad 2.66 \times 10^{-3} eV \leq \langle m_\nu\rangle \leq 2.86 \times 10^{-3} eV$$

$$= 0.00 \Rightarrow \quad \langle m_\nu\rangle = 2.76 \times 10^{-3} eV$$

$\langle m_\nu\rangle$ *from inverted mass hierarchy*

In this case, $m_3 \ll m_1 < m_2$ and $m_3 \ll \sqrt{\Delta m_A^2}$ so that from Eqn. (24) and (25), we have

$$m_1 \approx \sqrt{\Delta m_A^2}\ , \ m_2 \approx \sqrt{\Delta m_A^2}$$

The effective mass of neutrino $\langle m_\nu\rangle$ is then given by Eqn. (26) as

$$|\langle m_\nu\rangle| \approx \sqrt{\Delta m_A^2}\left|U_{e1}^2 + U_{e2}^2 e^{i\varphi 2}\right| \tag{33}$$

where we have neglected the small (< 5 %) corrections due to $|U_{e3}|^2$.

To obtain upper and lower bound on $\langle m_v \rangle$ in the present case, we solve the above equation i.e..

$$|\langle m_v \rangle| \approx \sqrt{\Delta m_A^2} \left| \cos^2\theta_{13} \cos^2\theta_{12} + \cos^2\theta_{13} \sin^2\theta_{12} e^{1\varphi z} \right|$$

$$\langle m_v \rangle^2 \approx \Delta m_A^2 [\cos^2\theta_{13} \cos^2\theta_{12} + \cos^2\theta_{13} \sin^2\theta_{12} e^{1\varphi 2}][\cos^2\theta_{13}$$

$$\cos^2\theta_{12} + \cos^2\theta_{13} \sin^2\theta_{12} e^{-i\varphi 2}] \approx \Delta m_A^2 (\cos^4\theta_{13})[(\cos^2\theta_{12} +$$

$$\sin^2\theta_{12})^2 - 2\cos^2\theta_{12} \sin^2\theta_{12}(1-\cos\varphi_2)] \Rightarrow \langle m_v \rangle \approx \sqrt{\Delta m_A^2}$$

$$(\cos^2\theta_{13}) \left[1 - \sin^2 2\theta_{12} \sin^2 \left(\varphi_2 / 2 \right) \right]^{1/2} \quad (34)$$

From Eqn. (34), the upper and lower bound on $\langle m_v \rangle$ can be derived as

$$\langle m_\downarrow v \rangle_\downarrow \max{}^\uparrow IH \approx \cos^2\theta_{13} \sqrt{\Delta m_A^2}$$

and

$$\langle m_\downarrow v \rangle_\downarrow \min{}^\uparrow IH \approx \cos 2\theta_{12} \cos^2\theta_{13} \sqrt{\Delta m_A^2} \quad (35)$$

corresponding to $\varphi_2 = 0$ and π respectively.

Putting the value of θ_{12} and θ_{13} from Eqn. (31) in Eqn. (35), we get following bound on $\langle m_v \rangle$

$$\langle m_\downarrow v \rangle_\downarrow \max{}^\uparrow IH \approx 4.74 \times 10^{-2} eV$$

and

$$\langle m_\downarrow v \rangle_\downarrow \max{}^\uparrow IH \approx 1.74 \times 10^{-2} eV \quad (36)$$

$\langle m_v v \rangle$ from almost Degenerate (A.D.) mass hierarchy

In case of normal and inverted hierarchy mass spectra, the lightest neutrino mass was assumed to be small. The existing bound on the absolute value of neutrino mass given by Mainz and Troitsk experiments is:

$$m_1 = 2.2 \text{ eV}$$

while the future tritium experiment KATRIN is planned to achieve the sensitivity $m_1 \approx 0.25$ eV.

The Wilkinson Microwave Anisotropy Probe (WMAP) and 2 degree Field Galaxy Redshift Survey (2dFGRS) data implies.

$$\sum_i mi \leq 0.7eV \tag{37}$$

Also, from the analysis of the latest Sloan Digital Survey Data and WMAP data, it was found that

$$\sum_i mi \leq 1.7eV$$

which for the case of three massive neutrinos implies

$$m_l = 0.6\ eV \tag{38}$$

These experimental bounds do not exclude the possibility that the lightest neutrino mass is much larger than $\sqrt{\Delta m_A^2}$. Hence, we may assume: $m_1 \approx m_2 \approx m_3$

so that the effective Majorana mass is given by

$$|\langle M_\nu \rangle| \approx m_1 \left| \sum_{i-1,2} U_{ei}^2 \right| \tag{39}$$

Neglecting a small contribution due to parameter $|U_{e3}^2|$ and following Eqn. (27), we get

$$\langle m_\nu \rangle \approx m_1 \cos^2 \theta_{13} \left[1 - \sin^2 2\theta_{12} \sin^2 \left(\varphi_2 / 2 \right) \right]^{1/2}$$

Using the value of θ_{12} and θ_{13}, we obtain the bound on $\langle m_\nu \rangle$ in case of A.D. hierarchy as

$$\langle m_{\downarrow}v\rangle_{\downarrow}\ \max^{\uparrow} AD \approx m_{\downarrow}1\cos^{\uparrow}2\theta_{\downarrow}13 \approx 0.98m_{\downarrow}1$$

and $$\langle m_{\downarrow}v\rangle_{\downarrow}\ \min^{\uparrow} AD \approx 0.36m_{\downarrow}1 \qquad (40)$$

As before, upper and lower limit corresponds to $\varphi_2 = 0$ and π respectively.

Taking $m_1 \approx$ 0.2 eV (sensitivity of tritium experiment KATRIN we obtain bound on $\langle m_v\rangle$ as

$$\langle m_{\downarrow}v\rangle_{\downarrow}\ \min^{\uparrow} AD \approx 0.72 \times [10]^{\uparrow}(-1)eV$$

$$\langle m_{\downarrow}v\rangle_{\downarrow}\ \max^{\uparrow} AD \approx 1.96 \times [10]^{\uparrow}(-1)eV \qquad (41)$$

Following the bounds obtained in Eqns. (32), (36) and (41) on effective Majorana mass of the neutrino, we have divided $\langle m_v\rangle$ in following three intervals-

(i) 1.67×10^{-3} eV $\leq \langle m_v\rangle^{NH} \leq 3.56\times10^{-3}$ eV Normal Hierarchy

(ii) 1.74×10^{-2} eV $\leq \langle m_v\rangle^{IH} \leq 4.74\times10^{-2}$ eV Inverted Hierarchy

(iii) 0.72×10^{-1} eV $\leq \langle m_v\rangle^{AD} \leq 1.96\times10^{-1}$ eV A.D. Hierarchy

Conclusions

The observation of the neutrino oscillations in the experiments with atmospheric, solar, reactor and accelerator neutrinos proves that neutrino masses are different from zero and that the states of flavor neutrinos v_e, v_u, v_t are mixtures of states of neutrinos with different masses. There are two different possibilities of neutrinos with definite masses: they can be Dirac particles possessing conserved lepton number which distinguishes neutrinos and antineutrinos or purely neutral Majorana particles with identical neutrinos and antineutrinos. It will be extremely important for the further development of the theory of neutrino masses and mixing to answer the fundamental question of neutrinos being Dirac or Majorana particles.

The neutrino oscillation experiments only provide the mass square difference of the eigen states at present so that absolute scale of neutrino mass can not be determined. However, it is expected that the following challenges can be addressed by future neutrino oscillation experiments:

1. ***The sign of*** Δm^2_{31}**:** whether the neutrino mass ordering is 'normal' or 'inverted' has important implications for Grand Unification, Flavour Models and Cosmology. This can be observed by noting the dependence of Δm^2_A on Δm^2_{31}
2. ***The question of CP-violation*** (δ)**:** measurement of the oscillation phase would signal CP violation in the lepton sector, which would also have profound implications for Grand Unification, Flavour Models and Cosmology.
3. ***High precision measurements of mixing angles*****:** especially reactor angle θ_{13} which has so far not been measured at all; a high precision determination of all the mixing angles provides crucial information for Grand Unification and Flavour Models.

References

1. **A.A. Aguilar-Arevalo *et al.*** [The MiniBooNE Collaboration], *Phys. Rev. Lett.* 98, 231801 (2007); hep-ex 0704.1500.
2. **S. N. Ahmed *et al.*** [SNO Collaboration]; nucl-ex/0309004.
3. **C. Athanassopoulos *et al.*** [LSND Collaboration], *Phys. Rev. Lett.* 81, 1774 (1998); nucl- ex/9709006.
4. **F.T. Avignone III, G.S. King III and Yu G. Zdesenko,** *New Journal of Physics* 7, 6 (2005).
5. **D.S. Ayres *et al.*** [NOíA Collaboration]; hep-ex/0503053.
6. **J. N. Bahcall,** arXiv: *physics*/0406040.
7. **C.L. Bennet *et al.***, astro-ph/0302207
8. **K. Eguchi *et al.*** [kamLAND Collaboration], Phy. Rev. Lett. 90, 021802 (2003): hep-ex/0212021.

9. **Y. Fukuda *et al.*** [super-kamiokande collaboration], *Phy. Rev. Lett.* 81, 1562 (1998).
10. **S. Geer,** *Phys. Rev.* D 57, 6989 (1998) [Erratum-ibid. D 59, 039903 (1999)]; hep-ph/9712290.
11. **Y. Itow *et al.*** [The T2K Collaboration]; hep-ex/0106019.
12. **Boris Kayser,** *The Physics of Massive Neutrinos*, World Scientific Publishing Co. Pte. Ltd., Singapore, 1989.
13. **S.F. King,** *physics.pop*-ph/0712.1750v1.
14. **V.M. Lobashev *et al.*, Phys. Lett.** B 460, 227 (1999); Nucl. Phys. Proc. Suppl. 91, 280 (2001).
15. **Z. Maki, M. Nakagawa and S.Sakata,** *Prog. Theo. Phys.* 28 (1962) 247; B. W. Lee, S. Pakvasa, R.E. Shrock and H. Sugawara, Phys. Rev. Lett. 38 (1977) 937 [Erratum-ibid. 38(1977) 1230].
16. **J. Maricic and J. G. Learned,** *Contemp. Phys.* 46, 1 (2005).
17. **Osipowicz *et al.*** [KATRIN Collaboration], hep-ex/0109033; V.M. Lobashev, Nucl. Phys. A 719, 153 (2003).
18. **Pontecorvo,** *Sov. Phy JETP* 7, 172 (1958) [Zh. Eksp. Teo r. Fiz 34, 247 (1957)].
19. **T. Schwetz,** hep-ph/0710.5027.
20. **T. Schwetz, M. Tortola and J.W.F. Valle,** arXiv: hep-ph/ 1103.0734v1.
21. **T. Schwetz, M.A.Tortola and J.W.F.Valle;** *New J. Phys.*, 10, 113011 (2008).
22. **M. Tegmark *et al.*,** astro-ph/0310723.
23. **Weinheimer *et al.*,** *Phys. Lett.* B 460, 219 (1999); hep-ex/0210050.
24. **L. Wolfenstein,** *Phys. Rev.* D17 (1978) 2369; S. Mikheyev and A. Yu. Smirnov, Sov. J. Nucl. Phys. 42, 913 (1985).

17

NUCLEAR DOUBLE BETA DECAY

K. Chaturvedi, Varun Yadav and Jai Prakash

Introduction

The four fundamental interactions known till data are the gravitation, the electromagnetism, the strong interaction and the weak interaction. The last one was observed about 110 years ago, in 1896 by H. Becquerel, who, at the suggestion of Poincare, was investigating the uranium salts to find out a relationship between the property of optical fluorescence and the newly discovered X-rays. In 1899, Rutherford discovered two types of radiations, which he called α-rays and β-rays. Villard, in 1909, discovered a third type of radiation called γ-rays. From further investigations, it is now well known that α-rays and β-rays are not so strictly radiations but corpuscles in nature and so they are described as α-particles and β-particles. The first known process of the weak interaction is the beta (β) decay. The β decay corresponds to the decay of a parent nucleus by the emission of an electron (e^-) and an anti neutrino (v_e) or a positron (e^+) and a neutrino (v), to any of the levels of the daughter nucleus through weak interaction. The $e^{\pm}$ are called β -particles. The history of neutrino started in 1930 with the proposal of Pauli in order to explain the continuous β-spectrum. He proposed that in the β-decay process, the electron is emitted together with a massless and chargeless particle of spin ½ He gave the name neutron (the neutral one) to this particle. The particle that today we call neutron, was discovered in 1932 by J. Chadwick in the nuclear reaction.

$$[\text{He}]^{\uparrow} 4 + [\text{Be}]^{\uparrow} 9(\text{C}^{\uparrow} 12 + \text{n} \quad (1)$$

After the discovery of neutron, the particle proposed by Pauli was called neutrino (little neutron) by Enrico Fermi. It is not before the work of Fermi that weak interaction received a well established theoretical fundament. Fermi, in 1934, gave the first successful theory of β - decay. Experimental detection of neutrino was done by F. Reines and C. W. Cowan at Savanah River experiment during 1954-1956. On June 14, 1956, Reines and Cowan sent a telegram to Pauli at Zurich University: *"We are happy to inform you that we have definitely detected neutrinos from fission fragments by obsessing inverse beta decay of protons. Observed cross section agrees well with expected six times ten to minus forty-four square centimeters."* In 1956, T. D. Lee and C. N. Yang, to resolve the $\tau - \theta$ puzzle, proposed for the first time that parity (P), which is a spatial symmetry, might not be conserved in weak interaction. In 1957 Mrs. Wu and co-workers of the National Bureau of Standards verified the hypothesis of Lee and Yang. The experiment seemed also to prove that the C symmetry, symmetry between particles and antiparticles was violated. In 1962, Leon M. Lederman, Melvin Schwartz and Jack Steinberger showed that more than one type of neutrino exists by first detecting interactions of the muon neutrino. When a third type lepton the tau (τ), was discovered in 1975 at the Stanford Linear Accelerator, it too was expected to have an associated neutrino. First evidence for this third neutrino type came from the observation of missing energy and momentum in τ decays analogous to the β-decay that had led to the discovery of the neutrino in the first place. The first detection of actual tau neutrino (v_t) interactions was announced in summer of 2000 by the DONUT collaboration at Fermilab.

The Neutrino

Neutrino is the most enigmatic member of the particle zoo. Neutrinos arc extremely difficult to detect because they hardly interact with matter at all - only one in a million of the neutrinos traversing Earth collide with an atomic nucleus. In 1967, Raymond

Davis, using a huge tank of cleaning fluid (HCl_4) placed deep underground to shield it from cosmic rays, started looking for neutrinos produced by the nuclear reactions in the Sun's core. However, he could only detect about half of what was expected. Other underground detectors found similar neutrino shortages. Bruno Pontecarvo in 1957 and independently, Maki, Nakagawa and Sakata in 1962, proposed that neutrinos might change from one flavor into another, a phenomenon called neutrino oscillation (õ-oscillation) Gradually, physicists started to suspect that oscillation could explain the shortfall of neutrinos detected from the Sun. The detectors could capture only one type of neutrino, and neutrinos during their journey from the Sun changed flavor and escaped detection. According to the SM, neutrinos should have zero mass. However quantum theory requires that if neutrinos change flavor they must have mass. Several experiments with atmospheric neutrinos and so-called long-baseline experiments in which neutrinos are beamed from a source to a distant detector have confirmed that neutrinos change flavor and so have mass.

The Experimental Search for Neutrino MASS

There are two basic approaches for the experimental search of neutrino mass, namely terrestrial and extra-terrestrial approaches, which are discussed briefly in the below.

Terrestrial Approach

The experimental search under terrestrial approach can be further differentiated as accelerator and non-accelerator experiments.

Accelerator experiments: The accelerator experiments comprised of π, τ- decay and neutrino oscillations. Neutrino oscillations at accelerators are called appearance experiments because only those flavors appear at the end which were not present in the original beam. These appearance or the accelerator experiments are very sensitive to small mixing angles as in an ideal case, one expects a pure neutrino beam without any background. In accelerators, the conversion $\nu_\mu \rightarrow \nu_e$ and $\nu_\mu \rightarrow \nu_\tau$ is usually used.

Non-accelerator experiments: The non-accelerator experiments are the tritium β decay, neutrino oscillations at reactors, the neutrino decay at reactors and the $\beta\beta$ decay. In disappearance (reactor) experiments, less neutrinos of the same flavor reach the detector than would be expected from the source. The reactor experiments measure all channels e.g. $\nu_e \rightarrow \nu_\mu, \nu_\tau$ etc. However, these experiments require a precise knowledge of the neutrino flux which is a major disadvantage.

Extra-Terrestrial Approach

The most promising experiments in this category are investigations of solar neutrinos, atmospheric neutrinos, neutrinos from supernova explosions and the neutrino cosmic background radiation. Neutrino oscillation experiments only measure mass-square differences Δm^2 but do not provide information about the absolute neutrino spectrum and can not distinguish between pure Dirac and Majorana neutrinos. The atmospheric neutrino data in Super-Kamiokande experiment, confirms the indications of oscillations in earlier data from Kamiokande, IMB experiments. The five experiments namely Chlorine, Kamiokande, Gallex, SAGE and Super-Kamiokande experiments conducted at the Homestake mine, Kamioka in Japan, Gran Sasso in Italy and Baksan in Russia respectively also confirm the neutrino oscillation. Further, a deficit in the flux of neutrinos from the sun is observed as compared to the predictions of the standard solar model by Bahcall and his collaborators and by many other groups. Finally, the detection of separate neutrino flux from ν_e neutrinos through charged current reactions and the flux of all active neutrinos through the neutral current reactions at SNO suggested that the solar neutrino oscillation is inescapable.

The solar and atmospheric neutrino puzzle is explained by considering the neutrino oscillation among three neutrino species. The estimated neutrino mass-square differences are $\Delta m^2_{atm} \approx 2.0 \times 10^{-3}$ eV^2 and $\sin^2 2\theta_{atm} > 0.94$ for atmospheric neutrinos. The K2K long base line probes the ν_μ disappearance and provides first confirmation of atmospheric neutrino oscillation. In case of solar neutrinos, the $\Delta m^2_{sun} \approx 8.2\times10^{-5}$ eV² and $\sin^2 2\theta_{sun} > 0.82$. It has been further observed at the KamLAND reactor experiment that ν_e oscillates as well. The oscillation parameters are in agreement with those of solar neutrino experiments. This agreement based on CPT invariance shows that the matter effect on oscillation is well understood. The CHOOZ and Palo Verde reactor experiment data imply that the angle θ_{13} is extremely small. The cosmic background radiation and the study of galaxies can constrain the sum of the neutrino masse $\sum m_\nu \leq 25$ to 100 eV. Further, the cosmological data from WMAP and 2dFGRS give the constraint $\sum m_\nu \leq 0.7$ eV.

The limit on the mass of electron neutrino obtained from tritium β decay experiments of Mainz and Troitsk is $m_{\nu e} \leq 2.2$ eV. The proposed KARTIN experiment is expected to reach a limit of 0.25 eV. However, experimental study of tritium single β decay and β β decay together can provide sharpest limits on the mass and nature of electron neutrinos.

The Nuclear ββ Decay

The nuclear ββ decay is a rare second order weak transition between two isobars having even-even configuration and differ in nuclear change by two units. The ββ decay candidates are stable against single β decay either due to energy conservation or angular momentum mismatch. There are two modes of ββ decay, one involving the emission of two neutrinos (2vββ decay), which conserves the lepton number exactly and is an allowed process within SM, and the other neutrinoless double beta (0v ββ) decay, which violates the lepton number by twc units and has the potential to throw some light on physics beyond the SM. The neutrino emitting mode of the ββ decay can be regarded as a simultaneous transformation of two neutrons (bound in the initial nucleus) into two protons, which leads to the final state with emission of two electrons and two antielectron-neutrinos (2v ββ decay) i.e.

$$^{A}_{z}X \rightarrow ^{A}_{Z} +^{Y}_{2} + 2e^{-} + 2v_{e}^{-} \quad (2)$$

The three other ways through which 2v ββ decay may occur involving the emission (capture) of one or two positrons (electrons) along with two neutrinos are-

i. $^{A}_{z}X \rightarrow ^{A}_{Z} +^{Y}_{2} + 2e^{-} + 2v_{e}$ Positron emission (3)

ii $e^{\uparrow} - +A|zX(A|Z_{\downarrow}(-2)Y + e^{\uparrow} + +[2v]_{\downarrow} e$ Electron capture (4)

iii $2e^{-} +^{A}_{z} X \rightarrow ^{A}_{Z-2} Y + 2\upsilon_{e}$ i. Double electron capture (5)

The half-life of such decays was expected to be fairly long since they are low energy processes of second order in weak interaction. At the suggestion of Eugene P. Wigner, Maria Goepert-Mayer (1935) calculated the half-life for the 2v ββ decay to be > 10^{17} years, which is exceedingly slow even on the geological time scale.

The most interesting case of ββ decay, the 0 υ ββ decay was considered first by Racah in 1937 and W. Furry in 1939 as a tool to distinguish whether the neutrino is of Majorana (particle ≡ antiparticle) or Dirac (particle ≠ antiparticle) type. The mechanism of 0v ββ decay is based on the emission of an electron antineutrino $\overline{v}_{e}$ on the first decay vertex $(n \rightarrow p + e^{-} + \overline{v}_{e})$ and its absorption in the second vertex. The absorption part is inverse beta decay $(n + v_{\downarrow}e(p + e^{\uparrow} -)$ where electron neutrino instead of antineutrino is required. The emission and absorption processes yield to exchange of a virtual neutrino and the associated propagator produces a neutrino potential. Though the conservation of energy and momentum required the emission of a neutrino in single β decay, there is no corresponding requirement for neutrinos in 0*v* ββ. Energy and momentum could be conserved in this decay mode releasing two electrons only. Similar to the 2*v* ββ decay, the 0*v* ββ decay may also occur involving the emission (capture) of one or two positrons (electrons) as

(*i*) $^{A}_{Z}X \rightarrow ^{A}_{Z-2}{}^{Y} + 2e^{+}$ Positron emission (6) (*ii*) $e^- + {}^A_Z X \rightarrow {}^A_{Z-2} Y + e^+$ Electron capture (7)

(*iii*) $2e^- + {}^A_Z X \rightarrow {}^A_{Z-2} Y$ Double electron capture (8)

The first theoretical investigations of ββ decay were presented by Zeldovich et al., (1954) and Primakoff and Rosen. It was then realized that all ββ-decay enjoys a substantial kinematical advantage (~10^8) over the 2*ν* ββ-decay. This attracted the attention of experimentalists since life-times of about 10^{15} y were expected which were within reach of the then existing experimental capabilities. However, the 0 υ process was not observed even when this level was reached and one was tempted to interpret this as a manifestation that the neutrino is a Dirac particle. But, of course, this conclusion was premature. In fact the discovery of parity violation in weak interactions by Lee and Yang and Wu and coworkers and the fact that the neutrinos participating in weak interactions are always left handed, implied that the 0*ν* ββ decay is helicity suppressed and perhaps unobservable if the neutrinos are very light even though they may very well be Majorana particles. This forced the experimentalists to be quite pessimistic and the experimental activity subsided. Later on, theoretically, the 0*ν* ββ decay was expected to proceed mainly via the small admixtures of right handed currents in the predominantly left handed Lagrangian and consequently, the experimental activity was resumed in the 1960's.

The 2 υ ββ Decay

The complexity of the calculation of the ββ decay matrix elements consists in the fact that it is a second order process; i.e., in addition to the initial and final nuclear states, knowledge of the complete set of states of the intermediate nucleus is required. In solving this problem, different models and nuclear structure scenarios have been applied. The study of 2*ν* ββ decay provides a nice way to check the validity of nuclear models employed so far for the nuclear structure calculations.

The inverse half-life of the 2ν $\beta\beta$ decay for the $0^+ \rightarrow 0^+$ transition is given by

$$\left[T_{\frac{1}{2}}^{2\nu}(0^+ \rightarrow 0^+)\right]^{-1} = G_{2\nu}|M_{2\nu}|_2 \qquad (9)$$

where $G_{2\nu}$ is the integrated kinematical factor and it can be calculated with good accuracy. The nuclear transition matrix element (NTME), $M_{2\nu}$, which is a model dependent quantity, is given by

$$M_{2\nu} = \sum_N \frac{\langle 0^+ \| \sigma\tau^+ \| 1_N^+ \rangle \langle 1_N^+ \| \sigma\tau^+ | 0^+ \rangle}{E_N - \frac{E_1 + E_F}{2}} \qquad (10)$$

The half-life of 2ν $\beta\beta$ decay has been measured for about 10 nuclei out of 35 possible candidates. Using the experimental half-life $T_{\frac{1}{2}}^{2\nu}$ and accurately known integrated kinematical factor, the values of $M_{2\nu}$ matrix element can be extracted directly from Eq. (9).

The 0ν $\beta\beta$ Decay

The $0\nu\beta\beta$ decay mode is far more interesting since it violates lepton number conservation by two units ($\Delta L = 2$) and is not allowed in SM. Further, the emission of a neutrino and its absorption as an antineutrino demands neutrino and its antineutrino to be same which is not possible in the standard electroweak SU(2) X U(l) model where the neutrino and its antineutrino are distinct particles. After the discovery of parity violation in weak interactions, an additional requirement arose that the Majorana neutrino emitted in the first neutron decay should reverse its helicity from right to left-handed, since otherwise it cannot be absorbed in the process of second neutron decay owing to its wrong helicity. Such a reversal might be caused by the neutrino mass (m_ν) and/or might occur explicitly through an admixture of right-handed current in weak interactions. The Majorana mass term requires the breaking of lepton number L. Since in general gauge theories the only gauge-anomaly free combination of quantum numbers is B-L, where B is the baryon number, we will talk about breaking of B-L symmetry. In the SM, the absence of the right-handed neutrinos and the existence of an exact global B - L symmetry guarantees that the neutrinos are mass less. The requirement of non-zero neutrino mass, being not fulfilled by SM, is contained quite naturally in theories beyond the SM. In the context of gauge theories, the possible ways for breaking the B - L symmetry are, explicit breaking of B - L symmetry where Lagrangian contains terms that break B - L symmetry, spontaneous breaking of local B - L symmetry and spontaneous breaking of global B - L symmetry.

Thus, the study of ββ decay in general and 0ν ββ decay in particular is a convenient tool to test the following important ramifications vis-a-vis constraints on parameters of various gauge theoretical models beyond the SM namely

- Lepton number violation,
- Mass and charge conjugation properties of the electron-neutrino and
- Possible right handed admixtures in the weak leptonic current.

Ongoing and Planned Experiments

The four principal nuclear ββ decay modes, namely (*i*) $2\nu\beta\beta$ (*ii*) $0\nu\beta\beta$ (*iii*) $0\nu\beta\beta\varphi$ and (*iv*) $0\nu\beta\beta\varphi\varphi$ decay (Fig. 1) are experimentally distinguishable through their spectral shapes.

They should occur universally in competition with each other, when kinematically allowed. The nuclear ββ decay is a second order weak process. The life-time is very long. A process with a half-life of the order of 10^{20} years must be detected in the presence of inevitable traces of radioisotopes with similar energy release but with decay rates faster by more than 10 orders of magnitude. These relatively active impurities particularly those from natural activity such as Uranium and Thorium decay series, artificially produced activity and those of cosmogenic origin are the serious sources of background. The broad generic looking spectrum of electron energies from the $2\nu\beta\beta$ decay offers little help in providing a distinctive signature. Therefore, it is not surprising that attempts to directly observe the $2\nu\beta\beta$ decay were unsuccessful for about 40 years.

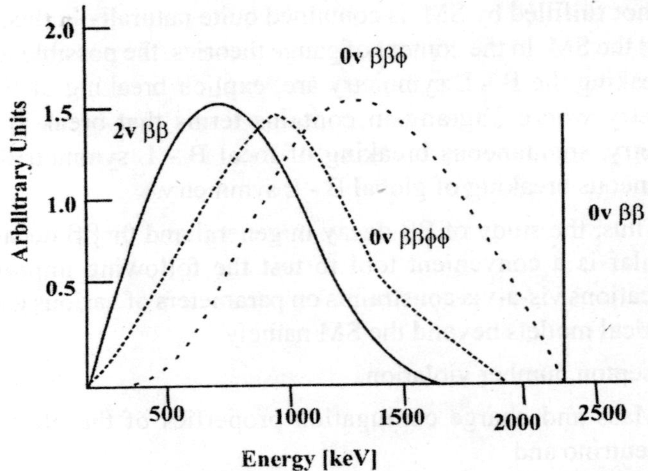

Fig. 1: Shapes of the Electron Sum Spectra in Cases of 2n ββ, 0n ββ, 0n βφ and 0n ββφφ Decay Modes.

The $2\nu\beta\beta$ decay has been already observed for 10 nuclei for the $0^+ \rightarrow 0^+$ transition and in the case of ^{100}Mo for the $0^+ \rightarrow 0_1^+$ transition. Further, limits on $2\nu\beta\beta$ decay for the $0^+ \rightarrow 2^+$ transition, $0\nu\ \beta\beta$ decay, $0\nu\beta\beta\varphi$ and $0\nu\beta\beta\varphi\varphi$ decay for $0^+ \rightarrow 0^+$ transition have been extracted. The observed half-lives of $2\nu\beta\beta$ decay and half-live limits for the $0\nu\beta\beta$ decay are 10^{19} to 10^{24} yr and $\geq 10^{25}$ yr respectively. Klapdor and his group have reported that the $0\nu\beta\beta$ decay has been observed in 76Ge. The results are controversial but it is expected that the issue will be settled soon [Klapdor et al. (2002), (2004) and Aalseth et al. (2002)].

The aim of all the present experimental activities is to observe the $0\nu\beta\beta$ decay. A number of review articles have been published over the years to up date the experimental status of the $\beta\beta$ decay.

Experiments on ββ decay can be divided into two broad categories namely

(*a*) indirect experiments

(*b*) direct experiments.

Indirect Experiments

Under indirect methods, the geochemical and the radiochemical methods are uncontrolled and controlled inclusive studies respectively in which neither the decay processes nor the final states are identified. We outline briefly the scope of different experimental methods in the below.

Geochemical Method

In geochemical experiments, one measures the abundance of daughter isotope that has accumulated over the geological time of the order of 10^9 years in an ore sample rich in parent isotopes. In such indirect experiments, the investigation consists of

- The determination of the age of the sample if possible by more than one independent methods.
- The extraction of the daughter nuclei from the sample and the determination of their concentration. This limits the applicability of the method to these cases where the daughter nucleus is a noble gas. There exists in nature only three systems namely ^{82}Se → ^{82}Kr and ^{128,130}Te → ^{128,130}Xe.
- A mass spectroscopic isotopic analysis of the extracted gases and determination of the absolute concentration of ^{82}Kr and ^{128,130}Xe isotopes.
- The determination of the coefficient of retention of the daughter nuclei in the sample during the time of their generation. Ideally, the system must be closed for both the parent and the daughter nucleus.
- A comprehensive and careful elimination of various background reactions, which may populate the daughter nucleus.

The signature for the production of a daughter isotope in ββ decay is an increase in the isotopic abundance relative to some reference isotopes of the same element not produced in ββ decay in the parent ore. State-of-the-art technique in mass spectrometry permits one to measure changes in the isotopic abundance of one part in 105 at the best. Thus, for accumulation time of 10^9 years and typical ββ decay rate of 10^{-21} events per year, the initial concentration of the daughter isotope in the ore sample must be no more than 10^{-7} that of the parent. Essentially, this requirement limits studies to transitions involving noble gas daughter isotopes, as only in those cases is such strong fractionation between elements with $|\Delta Z| = 2$ likely. Of course, the three reactions that produce daughter nuclei not found naturally (^{146}Sm, ^{232}U and ^{238}Pu) are in principle exceptions to the general rule. However, the latter two isotopes are so short-lived (∀10^2 years) that they may be better subjects for radiochemical methods while the ββ reaction producing the first isotope is enormously suppressed by phase space (T_0 = 61 keV). Manuel has reviewed in detail the geochemical methods and obtained experimental results.

The inherent limitation of the geochemical method is that it can not distinguish the 0*ν* ββ and the 2*ν* ββ decay modes, as the energy information is long lost. However, information about the decay modes can be sometimes inferred from the geochemically measured half-life by resorting to a specific nuclear model. For example, when the geochemical value is shorter than the lower limits for the 0*ν*ββ and Majoron accompanied 0*ν*ββ decay half-lives determined for that isotope by an independent technique, a predominance of the 2*ν*ββ decay mode is implied. The latter can be used to set limits on the neutrino mass and Majoron coupling strength as the half-life of any mode can not be shorter than the geochemical value.

Radiochemical Methods

Radiochemical methods are another version of indirect experiments in which the accumulation of the daughter nucleus occurs in a known interval of time under carefully controlled environment. The sample should be purged of daughter atoms existing in the environment as the accumulation time is much shorter than the geological time. Hence, the most easily studied reactions by the radiochemical method are $^{232}Th \rightarrow {}^{232}U$, $^{238}U \rightarrow {}^{238}Pu$ and $^{244}Pu \rightarrow {}^{244}Cm$ as the daughter isotopes are so short lived ($\approx 10^2$ years) that their abundance in the environment will be pretty low. The development of extremely sensitive resonance ionization techniques for noble gases suggest that radiochemical measurements of $\beta\beta$ decay for ^{82}Se and $^{128,130}Te$ may be practical. In such single atom experiments, it must be established that the container holding the parent is free from the contamination of the daughter present in the environment.

Direct Detection Experiments

The direct methods began in the late 1940's at approximately of the same time as the geochemical experiments. Various approaches were attempted such as coincidence searches for two emitted electrons with Geiger tubes and scintillators, nuclear emulsion techniques, magnetic spectrometers, tracking chambers and calorimeters using Ge crystals and ionization chambers. An excellent historical review of all these unsuccessful attempts has been presented by Haxton.

The following conditions are necessary requirements for an effective experimental set up.

- The source, detector and shielding should be free from radioactive contaminants at the level of one part per 10? for maximum background suppression.
- The reliable functioning of the equipments over a long period of time is required.
- A maximum pure and large sample of optimum size has to be decided so that there is an increase in the number of counts without increasing the background beyond a certain level.
- Extraction of as many characteristics as possible of the $\beta\beta$ decay event such as tracking in addition to energy

measurement so that the process can be discriminated against the background.

There are two different types of direct ββ decay experiments namely (*a*) active source - in which the source and the detector are identical and (*b*) passive source - in which the source and detector are not identical. The active source methods are more sensitive for the 0í ââ decay mode and the passive source methods are the best suited for the 2*v* ββ mode. In both types of experiments, attempt is made to measure the characteristic properties of 0*v*ββ decay such as :

- Emission of two electrons from the same point inside the source
- Simultaneous emission of two electrons and
- Constancy of the energy of the two emitted electrons.

The experiments, which determine all the three characteristics are called "complete experiments" and those which measure only the second and the third are called "identification experiments".

The signature of 2*v*ββ decay is the emission of two electrons from a common point free from other activity for some hours before and after the event. With the development of the tracking capabilities of the time projection chamber (TPC), it was possible ultimately to identify the long sought ββ decay event separating it from the background observed in the 2*v*ββ decay of ^{82}Se. The 2*v* ββ decay now has been observed for nearly 10 nuclei namely ^{48}Ca, ^{76}Ge, ^{82}Se, ^{96}Zr, ^{100}Mo, ^{116}Cd, ^{128,130}Te, ^{150}Nd and ^{238}U and for the rest of the isotopes lower limits have been set on the 2*v* ββ decay.

The 0*v* ββ decay is not observed so far and all the present experimental activities are directed in this direction. As the broadly distributed sum energy spectra of 2*v* ββ, 0*v* ββ, 2*v*ββφ and 2*v*ββφφ yields to similar experimental tracks, the direct counting experiments reporting the 2*v*ββ decay also give lower limits of half-lives for corresponding processes. Most of the recent activities, however refers to direct counting experiments which measure the energy of the ββ emitted electrons and so the spectral shapes of the 2ν $\beta\beta$, $0\nu\beta\beta$ and $0\nu\beta\beta\varphi$ modes of decay. Some experimental devices track the electrons (and other charged particles) measuring the energy, angular distribution and topology of the events. The tracking capabilities are essential to discriminate the $\beta\beta$ signal from the background.

The experimental limit on $\langle m_{\nu}\rangle$ can be written as

$$\langle m_{\nu}\rangle = (2.50\times10^{-8}\,\text{eV})\left[\frac{W}{\int x\varepsilon F_N}\right]^{\frac{1}{2}}\left[\frac{B\Delta E}{MT}\right]^{\frac{1}{4}} \quad (11)$$

$$\langle m_{\nu}\rangle = (2.67\times10^{-8}\,\text{eV})\left[\frac{W}{\int x\varepsilon F_N}\right]^{\frac{1}{2}}\left[\frac{1}{\sqrt{MT}}\right] \quad (12)$$

where W is the molecular weight of the source material, f is the isotopic abundance, x is the number of $\beta\beta$ atoms per molecule, ε is the detector efficiency, B is the background (counts/ kg-yr-keV), ΔE is the energy window in keV around $Q\beta\beta$, M is the mass of the $\beta\beta$ emitter (kg), T is the time measurement in years and nuclear factor-of-merit F_N is given by $F_N = G_{0\nu}|M_{0\nu}|^2$ where $G_{0\nu}$ is the kinematical factor and the nuclear transition matrix element (NTME) $M_{0\nu}$, is the likeliness of the transition. The Eqs. (11) and (12) suggest that the requirements of effective experimental set up are

- A $\beta\beta$ source having large factor-of-merit F_N
- A large source mass
- A low back ground preferably almost zero background
- A high detector efficiency.

The types of detectors currently used are:

– Calorimeters in which the detector is also the $\beta\beta$ source (Ge diodes, scintillators-CaF2, CdWO4, thermal detectors, ionization chambers). They are calorimeters which measure the two electron sum energy and discriminate partially signals from the background by pulse shape analysis (PSD).

– Tracking detectors of the source ≠ detector type (TPC, drift chambers, electronic detectors). In this case, the ββ source plane(s) is placed within the detector tracking volume, defining two or more detector sectors.

– Tracking calorimeters: They are tracking devices where the tracking volume is also the ββ source. Only one of this type of device is operating (Xenon TPC), but there are others in project.

Calorimeters of good energy resolution and almost 100% efficiency (Ge detectors, bolometers) are well suited for $0\nu\beta\beta$ decay searches but they lack the tracking capabilities to identify the background on an event-by-event basis. PSD and simultaneous measure of heat and ionization does this work. The Monte-Carlo (MC) modeling of the background spectrum to be subtracted from the data is approximate, so radio impurities are first subtracted as much as possible and then tracing back MC model, the remaining contaminations are subtracted from the data. On the contrary, the identification capabilities of various types of chambers make them very well suited for $2\nu\beta\beta$ and $0\nu\beta\beta$ φ searches. However, their energy resolution is rather modest and efficiency is only of a few present. Furthermore, the ultimate major background source in these devices when looking for the $0\nu\beta\beta$ decay will be that due to standard $2\nu\beta\beta$ decay. The rejection of the background provided by tracking compensates the figure-of-merit in the search of $0\nu\beta\beta$ decay. The modular calorimeters can have large amount of ββ emitters. However, the current operating chambers- except the Xe/TPC- can not accommodate a large amount of 2β emitters in the source plate. Further, the tracking devices will have 10 kg. or more in future.

The examples of ββ emitters measured in direct counting experiments are ^{48}Ca, ^{76}Ge, ^{82}Se, ^{96}Zr, ^{100}Mo, ^{116}Cd, ^{130}Te, ^{136}Xe and ^{150}Nd. A number of ββ decay experiments running presently are Heidelberg-Moscow experiment (HM), IzGEX, MIBETA, UCI,

ELEGANTs, NEMO and Gotthard tunnel etc. A number of future planned projects are GENIUS, MAJORANA, MOON, CAMEO, DCBA, CUORE, EXO, CANDLES, GSO, COBRA, Xe136 and XMASS etc.

Theoretical Formalism of 0νββDecay

The effective weak interaction Hamiltonian density for β decay due to the W-boson exchange restricted to left handed currents is given by

$$H_W = \frac{G}{\sqrt{2}} j_{L\mu} J_L^{\mu} + \text{h.c.}, \qquad (13)$$

Where

$$\frac{G}{\sqrt{2}} = \frac{9^2}{8M_1^2}\left[\cos^2\varsigma + \left(\frac{M_1}{M_2}\right)^2 \sin^2\varsigma\right] \qquad (14)$$

The gauge bosons W_L and W_R are the superposition of mass eigenstates W_1 and W_2 with masses M_1 and M_2, respectively and mixing angle ς, and $G = 1.16637 \times 10^5\ GeV^2$. The left handed V – A weak leptonic charged current is given by

$$[j]_{\downarrow} L\mu = c^{-} \gamma_{\downarrow}\mu([1-\gamma]_{\downarrow} 5)(_{\downarrow} eL \qquad (15)$$

where

$$(_{eL} = \sum_{\iota=1}^{2n} U_{ei} N_{iL} \qquad (16)$$

Here, N_i is a Majorana neutrino field with mass m_i in Eq (16), a Dirac neutrino is expressed as a superposition of a pair of mass degenerate Majorana neutrinos in the most general form. Further, the mixing parameters are constrained by the following orthonormality conditions:

$$\sum_{\iota=1}^{2n} |U_{ei}|^2 = \sum_{i=1}^{2n} |V_{ei}|^2 = 1 \text{ and} \sum_{i=1}^{2n} U_{ei} V_{ei} = 0 \qquad (17)$$

The strangeness conserving V-A hadronic currents are given by –

$$J_L^{\mu t} = g_V \bar{u}\gamma^{\mu}(1-\gamma_5)d \tag{18}$$

where

$$g_V = \cos\theta_c \tag{19}$$

In Eq. (19), the θ_c is the cabibbo-Kobayashi-Maskawa (CKM) mixing angle for the left handed d and s quarks.

In the mass mechanism, under light neutrino approximation, the inverse half life of $(\beta\beta)_0 v$ decay in the 2n mechanism for the $0^+ \rightarrow 0^+$ transition is given by –

$$[T_{\downarrow}(1/2)^{\uparrow}0((0^{\uparrow}+ \rightarrow 0^{\uparrow}+)] = ((\langle m_{\downarrow}()/m_{\downarrow}e)^{\uparrow}2G_{\downarrow}01(M_{\downarrow}GT - M_{\downarrow}F)^{\uparrow}2$$

$$= ((\langle m_{\downarrow}()/m_{\downarrow}e)^{\uparrow}2G_{\downarrow}01\left|M^{\uparrow}0v\right|^{\uparrow}2 \tag{20}$$

Where

$$\langle m_{\downarrow}() = \sum{}_{\downarrow}i^{\uparrow}\Box\ [U_{\downarrow}eu^{\uparrow}2_{\downarrow}i, m_{\downarrow}i < 10\text{MeV},] \tag{21}$$

In the closure approximation, NTMEs M^{0v} are written us –

$$\left|M^{0v}\right|^2 = \sum_{n,m}\langle 0_F^+\|0_{\mu,nm}\tau_n^+\tau_m^+\|0_i^+\rangle \tag{22}$$

where the nuclear transition operators are given by –

$$0_F = \left(\frac{g_V}{g_A}\right)^2 H_m(r)$$

&

$$O_{GT} = \sigma_1.\sigma_2 H_m(r) \tag{23}$$

The neutrino potential $H_m(r)$ arising due to the exchange of light neutrinos is defined as –

$$H_m(r) = \frac{4\pi R}{(2\pi)^3}\int d^3q\frac{\exp(iq.r)}{\omega(\omega+\bar{A})}$$

$$= \frac{2R}{\pi r}\int\frac{\sin(qr)}{(\omega+\bar{A})}dq$$

$$= \frac{R}{r}\phi(\bar{A}r) \tag{24}$$

In going from the first to the second line of Eq. (24) the neutrino mass has been neglected in comparison with the typical neutrino momentum $q \cong 200\ m_e$, Further

$$\bar{A} = \langle E_N \rangle - \frac{1}{2}(E_I + E_F) \tag{25}$$

The function $\phi(x)$ is detined by –

$$\phi(x) = \frac{2}{\pi}[\sin[x ci(x) - \cos x si(x)]] \tag{26}$$

Where

$$ci(x) = -\int_x^{\infty} t^{-1} \cos t dt$$

and

$$si(x) = -\int_x^{\infty} t^{-1} \sin t dt \tag{27}$$

The nuclear transition matrix elements (NTMEs) are calculated in PHFB model by employing the HFB wave functions.

The PHFB Model

The nuclear many-body Hamiltonian is written as where $V=V_{\{eff\}}$. In the HFB theory, one considers the general Bogoliubov transformations to define the quasiparticle creation and destruction operators.

$$H = \sum_{\alpha} \varepsilon_{\alpha} a_{\alpha}^{\dagger} \alpha\beta + \frac{1}{4}\sum_{\alpha\beta\gamma\delta} \langle \alpha\beta|V|\gamma\delta \rangle a_{\alpha}^{\dagger} a_{\beta}^{\dagger} a\delta a\gamma \tag{28}$$

where $V=V_{\{eff\}}$. In the HFB theory, one considers the general Bogoliubov transformations to define the quasiparticle creation and destruction operators.

$$q^{\dagger}_{\alpha} = \sum_{\beta} (u_{\alpha\beta} a_{\beta}^{\dagger} + v_{\alpha\beta} a_{\beta})$$

$$q_{\alpha} = \sum_{\beta} (u^{*}_{\alpha\beta} a_{\beta}^{\dagger} + v^{*}_{\alpha\beta} a^{\dagger}_{\beta}) \tag{29}$$

The quasiparticle operators satisfy the fermion anti commutation relations given below.

$$[q_{\alpha}^{\dagger}, q_{\beta}] = \delta_{\alpha\beta;} \left[q_{\alpha}^{\dagger}, q_{\beta}^{\dagger}\right] = \left[q_{\alpha}, q_{\beta}\right] = 0$$

The transformation coefficients are real and satisfy appropriate orthonormality conditions. Further, the 2n×2n linear transformation

$$\begin{bmatrix} q^{\dagger} \\ q \end{bmatrix} = \begin{bmatrix} u & v \\ v^{*} & u^{*} \end{bmatrix} \begin{bmatrix} a^{\dagger} \\ a \end{bmatrix}$$

is required to be unitary, and it follows from the unitarity condition that

$$uu^{\dagger} + vv^{\dagger} = u^{\dagger}u + \hat{v}v^{*} = 1$$

$$u\hat{v} + v\hat{u} = u^{\dagger}v + \hat{v}u^{*} = 0 \tag{30}$$

The quasiparticle vacuum $|\Phi_0\rangle$ is defined through the relation

$$q_{\alpha}|\Phi_0\rangle = 0$$

or

$$|\phi_0\rangle = 0 = (\text{Normalisation})\Pi_{\alpha} q_{\alpha}|0\rangle$$

where $|0$ is the particle vacuum and the normalization constant has to be appropriately inserted. The quasiparticle transformations do not conserve the number of particles. Hence, while minimizing the complete Hamiltonian H, we impose constraints via the use of Lagrange multipliers such that the average number of nucleons in the state $|\Phi_0\rangle$ be equal to the given number of nucleons of the nucleus under consideration. That is, we minimize

$$H' = H - \lambda_{\pi}N_{\pi} - \lambda_{\nu}N_{\nu} \tag{31}$$

Where

$$N_{\pi} = \sum_{\alpha} a^{\dagger}_{\alpha\pi} a_{\alpha\pi} \text{ and } N_{\nu} \sum_{\alpha} a^{\dagger}_{\alpha\nu} a_{\alpha\nu}$$

possess the expectation values

$$\langle\phi^0|N_{\pi}|\phi^0\rangle = Z, \langle\phi^0|N_{\nu}|\phi^0\rangle = A - Z \tag{32}$$

Further, the Lagrange's multipliers λ_{π} and λ_{ν} turn out to have the physical interpretation of proton and neutron fermi energies.

In the next step, we normal order the products of operators appearing in H' using Wick's theorem. The many-body Hamiltonian can be written with respect to the quasiparticle vacuum as

$$H' = H^{0'} + H^{2'} + H^{4'} \tag{33}$$

Where

$$H_0' = \mathrm{Tr}\left[\left(T - \lambda_\pi N_\pi - \lambda_\nu N_\nu + \frac{1}{2}\Gamma\right)\rho + \frac{1}{2}\Delta t^\dagger\right]$$

$$H_2' = \sum_{\alpha\beta}(h - \lambda_\pi N_\pi - \lambda_\nu N_\nu)_{\alpha\beta} : a_\alpha^\dagger a_\beta : + \frac{1}{2}\sum [\Delta_{\alpha\beta} : a_\alpha^\dagger a_\beta^\dagger : + \Delta_{\alpha\beta}^\dagger : a_\alpha a_\beta :]$$

$$H_4' = \frac{1}{4}\sum_{\alpha\beta\gamma\delta}\langle\alpha\beta|V|\gamma\delta\rangle : a_\alpha^\dagger a_\beta^\dagger a_\delta a_\gamma : \tag{34}$$

Here, the HF Hamiltonian h, the HF potential Γ and the pair potential Δ are given as

$$h = T + \Gamma$$

$$\Gamma_{\alpha\beta} = \sum \langle\alpha\gamma|V|\beta\delta\rangle_{\rho_{\delta\gamma}}$$

$$\Delta_{\alpha\beta} = \frac{1}{2}\sum_{\gamma\delta}\langle\alpha\beta|V|\gamma\delta\rangle t_{\gamma\delta} \tag{35}$$

The density matrix ρ and the pairing potential t are defined as follows

$$\rho_{\alpha\beta} = \langle\Phi_0|a_\beta^\dagger a_\alpha|\Phi_0'\rangle$$

$$= \sum_\delta v_{\alpha\beta} v_{\delta\alpha}^* = (v^\dagger v)_{\alpha\beta} \tag{36}$$

$$t_{\alpha\beta} = \langle\Phi_0|a_\beta a_\alpha|\Phi_0\rangle = v^\dagger u \tag{37}$$

For any normal product :0:, the expectation value

$$\langle\phi_0| : 0 : |\phi_0\rangle = 0$$

by construction. Thus, we get

$$\langle\phi_0|H_2' + H_4'|\phi_0\rangle = 0$$

and

$$\langle\phi_0|H'|\phi_0\rangle = \langle\phi_0|H_0'|\phi_0\rangle$$

$$= \mathrm{Tr}\left[\left(T - \lambda_\pi N_\pi - \lambda_\nu N_\nu + \frac{1}{2}\Gamma\right)\rho + \frac{1}{2}\Delta t^\dagger\right] \qquad (38)$$

The first term on the right hand side is the HF binding energy and the second term is the pairing ener-gy. The quasiparticle excitation energies are obtained from H_2' and the remaining H_4' contains the quasiparticle interaction energies, which are neglected in the HFB theory. To summarize, the HFB theory is a generalized HF procedure in which the HF field and the pairing interaction are treated simultaneously and on equal footing. There are several equivalent methods for deriving the HFB equations such as (*i*) variational procedure, (*ii*) the equation of motion and (*iii*) elimination of dangerous diagrams.

Henceforward, we use time reversal symmetry as we are concerned with even-even nuclei only. The HFB equations are obtained by equating to zero the off diagonal bilinear quasiparticle part of the transformed Hamiltonian after making a canonical transformation from a_a's to quasi-fermions q_a's through

$$q_k^\dagger = \sum_\alpha (u_{k\alpha} a_\alpha^\dagger + v_{k\alpha} a_{\bar{\alpha}})$$

$$q_k^\dagger = \sum_\alpha (u_{k\alpha} a_\alpha^\dagger - v_{k\alpha} a_\alpha) \qquad (39)$$

The HFB equations are written as

$$E_k u_{ka} = \sum_\gamma h'_{\alpha\gamma} u_{\alpha\gamma} + \Delta_{\alpha\bar{\gamma}} v_{\alpha\gamma}$$

$$E_k u_{ka} = \sum_\gamma \Delta_{\alpha\bar{\gamma}} u_{\alpha\gamma} - h'_{\alpha\bar{\gamma}} v_{\alpha\gamma} \qquad (40)$$

where E_k is the quasiparticle energy. Further

$$h'_{a\gamma} = |T_{\alpha\gamma} - \lambda\delta_{\alpha\gamma}\Gamma_{\alpha\gamma} \qquad (41)$$

Where

$$\Gamma_{\alpha\gamma} = \sum_{\beta\delta} [\langle\alpha\beta|V|\gamma\delta\rangle + \langle|\alpha\bar{\beta}|V|\gamma\bar{\delta}\rangle]\rho_{\beta\delta}$$

and

$$\rho_{\beta\delta} = \sum_k \nu_{k\beta}\nu_{k\delta}$$

The coupled HFB equations can be written in a simplified form making use of the Bloch-Messiah theorem. We define transformations

$$b_k^{\dagger} = \sum_{\alpha} C_{k,a} a_{\alpha}^{\dagger}$$

and

$$b_{\bar{k}}^{\dagger} = \sum_{\alpha} C^{*}_{k,a} a_{\alpha}^{\dagger} \tag{42}$$

where the expansion coefficients appearing in Eq. (42) can be obtained by diagonalizing the HF like potential h' in spherical basis which includes the appropriate density pk = v_{k2}, where

$$h'_{\alpha\beta} = \langle\alpha|T - \lambda_{\pi}N_{\pi} - \lambda_{\nu}N_{\nu}|\beta\rangle + \sum_k \langle ak|V|\beta k\rangle v_k^2 \tag{43}$$

The occupation probabilities vk^2 are obtained by solving the BCS equation

$$\Delta_k = \Delta_{kk'} = \sum_{k'} \langle k\bar{k}|V|k'\bar{k}'\rangle u_k{}' v_k{}' \tag{44}$$

The calculation involves iteration between Eqs. (43) and (44) until a reasonable convergence is achieved in terms of both the expansion coefficients C_{ka} and vk^2.

The ground state energy E_{HFB} is given by

$$E_{HFB} = \sum_{k=1}^{n} (T_{kk} + \lambda - E_k)v_k^2$$

The axially symmetric HFB intrinsic state with K=0 can be written as

$$|\Phi_0\rangle = \prod_{im} (u_{im} + v_{im} b_{im}^{\dagger} b_{\overline{im}}^{\dagger}|0\rangle \tag{45}$$

where $b_{im}^{\dagger}$ & $b_{\overline{im}}^{\dagger}$ are the creation operators.

The wave function $|\Phi_0\rangle$ can be recast into the form

$$|\Phi_0\rangle = N\exp\left(\frac{1}{2}\right)\sum_{\alpha\beta}\int_{\alpha\beta} a_\alpha^\dagger a_\beta^\dagger |0\rangle \tag{46}$$

Here N is a normalization constant. Using the standard projection technique, a state with good angular momentum is obtained from the HFB intrinsic state through the following relation.

$$|\Psi_0^J\rangle = p_{00}^J|\Phi_0\rangle$$

$$= \left[\frac{(2J+1)}{8\pi^2}\right]\int D_{00}^J(\Omega)\Phi_0\rangle d\Omega \tag{47}$$

$\langle m_\nu \rangle$ Using Experimental Half Lives From 0*ν* ββ Decay

We have calculated effective Majorana mass of light neutrino $\langle m_\nu \rangle$ in mass mechanism for the $0^+ \rightarrow 0^+$ transition, from Eqn. (20) using the experimentally available half lives $T^{0\nu}_{1/2}$ and theoretically calculated nuclear transition matrix elements (NTMEs) $|M^{0\nu}|$ for ^{96}Zr, ^{100}Mo, ^{130}Te, ^{150}Nd isotopes and 116Cd, 136Xe isotopes.

We have given experimental half-lives for above isotopes in Table 1. Further, the NTMEs $|M^{0\nu}|$ and calculated values of $\langle m_\nu \rangle$ are presented for above mentioned nuclei in Table 2.

Table 1: Experimental Half Lives for ^{96}Zr, ^{100}Mo, ^{116}Cd, ^{130}Te, ^{136}Xe and ^{150}Nd Isotopes

Nuclei	Experiment	$T^{0\nu}_{\frac{1}{2}}$ (Yrs)	Reference
^{96}Zr	NEMO-3	$>9.2 \times 10^{21}$	Argyriades *et al* (2010)
^{100}Mo	NEMO-3	$>1.0 \times 10^{24}$	Barabash *et al* (2010)
^{116}Cd	SOLOTVINO	$>1.7 \times 10^{23}$	Danevich *et al* (2003)
^{130}Te	CUORICINO	$>2.8 \times 10^{24}$	Arnaboldi *et al* (2008)
^{136}Xe	DAMA	$>5.0 \times 10^{23}$	Bernabei *et al* (2002)
^{150}Nd	NEMO-3	$>1.8 \times 10^{22}$	Argyriades *et al* (2009)

Table 2: Nuclear Transition Matrix Elements (NTMEs) $|M^{0\nu}|$ of ^{96}Zr, ^{100}Mo, ^{116}Cd, ^{130}Te, ^{136}Xe and ^{150}Nd Isotopes Along with Calculated Values of Effective Neutrino mass $<m_\nu>$

Nuclei	$\lvert M^0n \rvert$	Ref.	$<m_\nu>$(eV)	
			Min	Max
$^{100}Mo \rightarrow {}^{100}Ru$	4.71-7.77	Rath *et al* (2010)	0.20	0.32
$^{116}Cd \rightarrow {}^{116}Sn$	2.51-4.52	Simkovic *et al* (2009)	1.28	2.30
$^{130}Te \rightarrow {}^{130}Xe$	2.99-5.12	Rath *et al* (2010)	0.29	0.51
$^{136}Xe \rightarrow {}^{136}Ba$	1.71-3.53	Simkovic *et al* (2009)	0.99	2.04
$^{150}Nd \rightarrow {}^{150}Sm$	1.98-3.70	Rath *et al* (2010)	2.35	4.39

In Table 3, we have presented the predicted half lives due to normal, inverted and degenerate mass hierarchies along with experimentally available half lives.

Effect of Mixing Angles on $\langle m_\nu \rangle$

It is seen that value of effective neutrino mass strongly depends upon mixing angles θ_{12} and θ_{13}. Hence we have tried to show the effect of solar mixing angle θ_{12} on $\langle m_\nu \rangle$ by varying this angle (keeping θ_{13} fixed) and obtaining neutrino mass. The results are shown in Table 4.

Results and Discussions

In Table 1, experimental half lives for ^{96}Zr, ^{100}Mo, ^{116}Cd, ^{130}Te, ^{136}Xe and ^{150}Nd isotopes are shown. These data are taken from various ongoing DBD experiments like NEMO-3, CUORICINO, DAMA etc. In Table 2, we have given nuclear transition matrix elements (NTMEs) $|M^0n|$ of ^{96}Zr, ^{100}Mo, ^{116}Cd, ^{130}Te, ^{136}Xe and ^{150}Nd isotopes which are model dependent and are calculated in PHFB and QRPA models. These matrix elements are further employed to calculate the effective neutrino mass $\langle m_\nu \rangle$ for above isotopes by taking the experimental values of $T_{\frac{1}{2}}^{0\nu}$, as shown in Table 1. From

Table 3: Experimental and Predicted Half Lives (yrs.) of ^{96}Zr, ^{100}Mo, ^{116}Cd, ^{130}Te, ^{136}Xe and ^{150}Nd Isotopes for Normal, Inverted and Almost Degenerate Hierarchy Cases. Only Lower Limits of $|M^0n|$ are Considered

Nuclei	Half-life $T^{0\nu}_{\left(\frac{1}{2}\right)}$	Half-life $T^{0\nu}_{\left(\frac{1}{2}\right)}$ (predicted)		
	(Exp.)	Normal Hierarchy	Inverted Hierarchy	Degenerate Hierarchy
		m_ν=1.67-3.56 meV (θ_{12}=34.2°, è$_{13}$=7.5°)	m_ν=17.4-47.4 meV (θ_{12}=34.2°, θ_{13}=8.1°)	m_ν=72-196 meV (θ_{12}=34.2°)
^{96}Zr	$>9.2\times10^{21}$	$(33.1-7.3)\times10^{28}$	$(30.5-4.1)\times10^{26}$	$(17.8-2.4)\times10^{25}$
^{100}Mo	$>1.0\times10^{24}$	$(37.7-8.3)\times10^{27}$	$(34.7-4.7)\times10^{25}$	$(20.3-2.7)\times10^{24}$
^{116}Cd	$>1.7\times10^{23}$	$(32.2-7.1)\times10^{28}$	$(29.6-4.0)\times10^{26}$	$(17.3-2.3)\times10^{25}$
^{130}Te	$>2.8\times10^{24}$	$(25.6-5.6)\times10^{28}$	$(23.6-3.2)\times10^{26}$	$(13.8-1.9)\times10^{25}$
^{136}Xe	$>5.0\times10^{23}$	$(74.3-16.3)\times10^{28}$	$(68.4-9.2)\times10^{26}$	$(39.9-5.4)\times10^{25}$
^{150}Nd	$>1.8\times10^{22}$	$(12.4-2.7)\times10^{28}$	$(11.5-1.5)\times10^{26}$	$(66.9-9.0)\times10^{24}$

Table 4: Variation of Effective Neutrino Mass $\langle m_\nu \rangle$ (eV) with Solar Mixing Angle θ_{12} for Three Types of Mass Hierarchies. The Theoretical Value of Neutrino Mass is also Given for Various DBD Isotopes

$\sin^2\theta_1$	Normal hierarchy		Inverted hierarchy		Almost degenerate hierarchy		Theoretical neutrino mass $\langle m_\nu \rangle$		
	$\langle m_\nu \rangle^{NH}_{min}$	$\langle m_\nu \rangle^{NH}_{max}$	$\langle m_\nu \rangle^{IH}_{min}$	$\langle m_\nu \rangle^{IH}_{max}$	$\langle m_\nu \rangle^{AD}_{min}$	$\langle m_\nu \rangle^{AD}_{max}$	Isotope	$\langle m_\nu \rangle_{min}$	$\langle m_\nu \rangle_{max}$
0.28	**1.57×10^{-3}**	3.24×10^{-3}	2.09×10^{-2}	4.74×10^{-2}	0.087	0.197	^{96}Zr	6.49	10.02
0.316	1.88×10^{-3}	3.55×10^{-3}	1.74×10^{-2}	4.74×10^{-2}	0.072	0.197	^{100}Mo	0.20	0.32
0.32	1.91×10^{-3}	3.59×10^{-3}	1.71×10^{-2}	4.74×10^{-2}	0.071	0.197	^{116}Cd	1.28	2.30
0.34	2.08×10^{-3}	3.76×10^{-3}	1.52×10^{-2}	4.74×10^{-2}	0.063	0.197	^{130}Te	0.29	0.51
0.36	2.25×10^{-3}	3.93×10^{-3}	1.33×10^{-2}	4.74×10^{-2}	0.055	0.197	^{136}Xe	0.99	2.04
0.38	2.42×10^{-3}	4.10×10^{-3}	1.14×10^{-2}	4.74×10^{-2}	0.047	0.197	^{150}Nd	2.35	4.39

table 2, the upper limit on neutrino mass comes out to be 0.20 eV corresponding to ^{100}Mo. In Table 3, we have presented the predicted half lives due to normal, inverted and degenerate mass hierarchies along with experimentally available half lives. From the table, it is clear that $T^{0\nu}_{\frac{1}{2}}$ due to inverted and degenerate hierarchies can be realized in future and ongoing experiments while $T^{0\nu}_{\frac{1}{2}}$ due to normal hierarchy of neutrino mass are far reaching in near future. Finally, in Table 4, we have shown the effect of solar mixing angle $\langle m_\nu \rangle$ θ_{12} on $\langle m_\nu \rangle$ by varying it in the range 32^0-38^0 keeping θ_{13} fixed. It is seen that $\langle m_\nu \rangle^{NH}_{min}$ and $\langle m_\downarrow \nu \rangle_\downarrow$ $\min^\uparrow$ *IH* depend strongly on θ_{12}. $\langle m_\downarrow \nu \rangle_\downarrow$ $\min^\uparrow$ *NH* goes on increasing while $\langle m_\downarrow \nu \rangle_\downarrow$ $\min^\uparrow$ *IH* decreases with increase in θ_{12}. The theoretically calculated values of neutrino mass $\langle m_\nu \rangle$ for various isotopes are also given. The minimum neutrino mass comes out to be 0.20 eV corresponding to ^{100}Mo. Out of three mass hierarchies, only degenerate one shows the comparative values of $\langle m_\nu \rangle$ with the theoretically predicted calculations while other two hierarchical values of $\langle m_\nu \rangle$ cannot be realized in near future.

Conclusions

Investigation of neutrinoless double beta (0ν $\beta\beta$) decay of nuclei is the only practical way which could allow to proof that neutrinos are Majorana particles. This is because there are huge numbers of parent nuclei in the source. However, if neutrinos are Majorana particles, possibility of $0\nu\beta\beta$ decay is extremely small. This is due to the fact that $0\nu\beta\beta$ decay is a second order process in Fermi constant and this decay mode is possible only due to neutrino helicity flip. Smallness of neutrino masses is the additional suppression factor in the decay probability.

Experiments on the measurement of the half-lives of such rare process as neutrinoless $\beta\beta$ decay with severe requirements to background and energy resolution are extremely difficult. A big progress is achieved. However, future experiments with about one ton detectors, which will allow to reach the region of the values of the effective Majorana mass, which is predicted from the neutrino oscillation data in the case of inverted hierarchy is definitely a challenge. Taking into account the importance of the problem of the nature of the massive neutrinos, there is no doubt that goals of future experiments will be achieved.

References

1. **C.E. Aalseth et al.,** *Nucl. Phys. B* (Proc. Suppl.) 48, 223 (1996); Phys. Rev. D 65, 092007 (2002); hep-ph/0202018.
2. **Q.R. Ahmed et al.,** *Phys. Rev. Lett.* 87, 071301 (2001); ibid 89, 011301 (2002); ibid 89, 011302 (2002).
3. **S.N. Ahmed et al.,** *Phys. Rev. Lett.* 92, 181301 (2004).
4. **M.H. Ahn et al.,** *Phys. Rev. Lett.* 90, 041801 (2002).
5. **J. Argyriades *et al.*** (NEMO Collaboration), *Phys. Rev.* C 80 (2009)032501, arXiv:0810.0248[hep-ex].
6. **J. Argyriades *et al.*,** *Nucl. Phys.* A 847 (2010)168–179, arXiv:0906.2694[nucl-ex].
7. **C. Arnaboldi *et al.*,** *Phys. Rev.* C 78 (2008) 035502, arXiv:0802.3439[hep-ex].
8. **F.T. Avignone III, G.S. King III and Yu. G Zdesenko,** *New Journal of Physics* 7, 6 (2005).
9. **S.Barabash, V.B. Brudanin, and T.N.Collaboration,** arXiv:1002.2862[nucl-ex].
10. **R. Barbieri, L.J. Hall, D. Smith, A. Strumia, and N. Weiner,** hep-ph/9807235.
11. **C.L. Bennet *et al.*,** astro-ph/0302207
12. **R. Becker-Szendy et al.,** *Nucl. Phys.* B (Proc. Suppl.) 38, 331 (1995).
13. **E. Bellotti et al.,** *Phys. Lett* 146 B, 450 (1984).
14. **R. Bernabei *et al.*,** *Physics Letters* B 546 (2002)23–28.
15. **C. Bloch and A. Messiah,** *Nucl. Phys.* 39, 95 (1962).
16. **F. Boehm et al.,** *Phys. Rev.* D 62, 072002 (2000).
17. **B.T. Cleveland et al.,** *Astrophys.* J 496, 505 (1998).

18. **F. A. Danevich *et al.*,** *Phys. Rev.* C 68 (2003) 035501.
19. **K. Eguchi et al.,** *Phys. Rev. Lett.* 90, 021802 (2003).
20. **H. Ejiri et al.,** *J. Phys.* G 17, 155 (1991); J. Phys. Soc. Jpn. 64, 339 (1995); Nucl. Phys. A 611, 85 (1996); Phys. Rev. Lett. 85, 245 (2000); Phys. Rev. C 63, 65501 (2001).
21. **S.R. Elliott, A.A. Hann, M.K. Moe,** *Phys. Rev. Lett.* 59, 2020 (1987); ibid 59, 1649 (1987); Nucl. Instrum. Methods A 273, 54 (1988).
22. **S. R. Elliott,** *Nucl*-ex/0312013.
23. **S.R. Elliott and J. Engel,** *J. Phys.* G: Nucl. Part. Phys. 30, R183 (2004).
24. **A. Faessler and F. Simkovic,** *J. Phys.* G 24, 2139 (1998); hep-ph/9901215.
25. **Fiorini,** *Phys. Rep.* 307, 309 (1998).
26. **E. Fiorini,** *Rev. Nuovo Cim.* 2, 1 (1972).
27. **Y. Fukuda et al.,** *Phys. Lett.* B 335, 237 (1994); Phys. Rev. Lett. 81, 1562 (1998); ibid 82, 1810 (1999); ibid 82, 2430 (1999); ibid 82, 2644 (1999); ibid 85, 3999 (2000); ibid 86, 5651 (2001); T. Okumura, Talk given at international Workshop on "Neutrino Oscillations in Venice", December 2003, Italy.
28. **W. Furry,** *Phys. Rev.* 56, 1184 (1939).
29. **M. Goeppert-Mayer,** *Phys. Rev.* 48, 512 (1935).
30. **W.C. Haxton and G.J. Stephenson Jr.,** *Prog. Part. Nucl. Phys.* 12, 409 (1984).
31. **G.S. Hurst, M.G. Payne,S. Kramer and C.H. Chen,** *Phys. Today*, 33, no. 9, 24 and Private Communication (1980).
32. **H.V. Klapdor-Kleingrothaus,** *Int. J. Mod. Phys.* A 13, 3953 (1998); hep-ex/9901021; Found. Phys. 33, 813 (2003).
33. **H.V. Klapdor-Kleingrothaus et al.,** *Mod. Phys. Lett.* A 16, 2409 (2001); hep-ph/0103062.
34. **H.V. Klapdor-Kleingrothaus and U. Sarkar,** hep-ph/ 0201224.
35. **H.V. Klapdor-Kleingrothaus, A. Dietz, H.L. Harney and I.V. Krivoshina,** hep-ph/0201231; hep-ph/0205228.
36. **V.M. Lobashev et al.,** *Phys. Lett.* B 460, 227 (1999); Nucl. Phys. Proc. Suppl. 91, 280 (2001).

37. **V.M. Lobashev,** *Nucl. Phys.* A719, 153 (2003).
38. **E. Majorana,** *Nuovo Cim.* 5, 171 (1937).
39. **O.K. Manuel,** *J. Phys.* G17, S221 (1991).
40. **M. Moe and P. Vogel,** *Ann. Rev. Nucl. Part. Sci.* 44, 247 (1994).
41. **R.N.Mohapatra and G.Senjanovic,** *Phys. Rev.* D 23, 165 (1981).
42. **R.N. Mohapatra and J.D. Vergados,** *Phys. Rev. Lett* 47, 1713 (1981).
43. **R.N. Mohapatra,** *Phys. Rev.* D 34, 909 (1986); ibid 34, 3457 (1986).
44. **R.N. Mohapatra, Neutrinos,Ed. H.V. Klapdor,** Springer-Verlog, 117 (1988).
45. **R.N. Mohapatra and P.B. Pal,** *Massive Neutrinos in Physics and Astophysics*, World Scientific, Singapore, (1991).
46. **R.N. Mohapatra,** *Unification and Supersymmetry* (Springer Heidelberg; New York) (1986) and (1992).
47. **N. Onishi and S. Yoshida,** *Nucl. Phys.* A260, 226 (1966), Nucl. Phys. 80, 367 (1966).
48. **A. Osipowicz et al.,** hep-ex/0109033.
49. **Pontecorvo,** *Sov. Phy JETP* 7, 172 (1958) [Zh. Eksp. Teo r. Fiz 34, 247 (1957)].
50. **H. Primakoff and S. P. Rosen,** *Proc. Phys. Soc.* (London) 78, 464 (1961); Phys. Rev. 184 1925 (1969).
51. **G. Racah,** *Nuovo Cim.* 4, 322 (1937).
52. **P.K. Rath, R. Chandra, K. Chaturvedi, P. K. Raina and J.G. Hirsch,** *Phys. Rev.* C 82, 064310 (2010).
53. **F. Simkovic *et al.*** *Phys. Rev.* C 79, 055501 (2009).
54. **T. Schwetz, M. Tortola and J.W.F. Valle,** arXiv: hep-ph/ 1103.0734v1.
55. **T. Schwetz, M.A.Tortola and J.W.F.Valle;** *New J. Phys.*, 10, 113011 (2008).
56. **F. Simkovic, P. Domin and S.V. Semenov,** nucl-th/0006084.
57. **D.N. Spergel et al.,** astro-ph/0302209.
58. **L.Weinheimer *et al.*,** *Phys. Lett.* B 460, 219 (1999); hep-ex/ 0210050.

18

SUPER HEAVY ELEMENTS (SHE)

K. Chaturvedi, Shweta Goswami and Omji Bajpai

Introduction

Super heavy elements commonly abbreviated as SHE are known as the transactinides elements, these elements have mass and atomic numbers beyond uranium, starting form Rutherfordium which has atomic number (Z=104). These elements are traditionally considered to be those which lie above elements Z= 103 the last of the actinides. Starting with Rutherfordium (Rf), element, these elements are sometimes referred to as the super-transactinides. Collectively, they represent the very top end of the periodic table of elements.

Due to the rapid increase of the repulsive Coulomb forces between the protons, the number of chemical elements is limited by fission. This macroscopic behavior is governed by shell effects, without which the nuclear chart may end near element 106. There is evidence to suggest that nuclei can survive beyond the macroscopic limit, far into the trans-uranium region, where the necessary balance between the nuclear force and the Coulomb force is achieved *only* through shell stabilization. SHE are hypothesized to exist near the next (predicted) double shell closure above Lead where they have may have surprisingly long half-lives, maybe even on the order of millions of years. This postulate has fuelled vigorous research in the field, thereby earning it the reputation of being a search for the next "magic" shell, a Holy Grail of contemporary physics.

Today's definition "super heavy" are nuclei with 'Z>103' (Fig. 1). All of the heaviest elements beyond plutonium are artificially produced in heavy-ion reactions. Once created, the chemistry and subsequent placement of heaviest elements in the periodic table can not be taken for granted. With an increasing number of nucleons, relativistic effects may influence the ordering of atomic orbitals and may play a progressively important role in defining the chemical properties of a given element. Super heavy elements are chemical elements which are much heavier than those which we know of from our daily life, are a persistent dream in human minds and the kernel of science fiction literature for about a century.

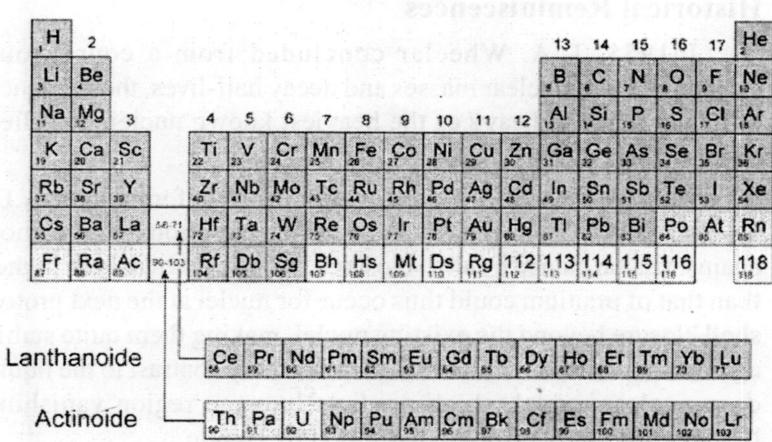

Fig. 1. Super Heavy Elements with Z>103

Verifying through independent chemical studies if element Z=112 belongs in group 12, remains a hot topic. In order to submit themselves to chemical studies, the nuclides must have half lives of at least a few seconds. All the observed isotopes of elements between Meitnerium (Mt, Z=109) and Roentgenium (Rg, Z=111) are too short lived. Chemical studies have been carried out for some Dubnium (Db, Z=105), Sg and Hassium (Hs, Z=108) isotopes and for the longer lived Z=112.

The nuclear shell model predicts that the next doubly magic shell closure beyond ^{208}Pb is at a proton number Z=114, 120 or 126 and at a neutron number N=172 or 184. The outstanding aim of experimental investigation is the exploration of this region of spherical 'Super heavy elements'. These heaviest elements are synthesized in accelerator based experiments with extremely high "in-flight" separation efficiencies ensuring that a single event of interest is separated from~10^{10}-10^{11} in beam products. Depending on the beam-target combination employed, two broadly defined categories of reaction mechanism are in use. The reaction products are typically created via successive alpha-decays ending by spontaneous fission in the region of the known elements.

Historical Reminiscences

In1955, J. A. Wheelar concluded from a courageous extrapolation of nuclear masses and decay half-lives, the existence of nuclei twice as heavy as the heaviest known nuclei, he called super heavy nuclei.

In 1966, in a study of nuclear masses and deformations, W.D. Myers and W.J. Swiatecki emphasized the enormous stabilization against fission gained by shell closures. Fission barriers even higher than that of uranium could thus occur for nuclei at the next proton shell closure beyond the existing nuclei, making them quite stable against spontaneous fission. This was in sharp contrast to the liquid drop nuclear model which predicts in same region vanishing barriers and hence, prompt disruption by fission.

Although the discussion focused on Z=126 as next proton shell closure, Z=114 was mentioned as an alternative with reference to unpublished calculations by H. Meldner, who presented his results later at the "why and how......" symposium in 1966, the seminal event for super heavy element research. A fantastic perspective was thus opened; an island of super heavy elements located not too far from the then heaviest known element 103 and hence perhaps within reach. First theoretical estimates of decay half lives around the doubly magic nucleus Z=114, N=184 revealed an island-like topology as shown below in Fig. 2.

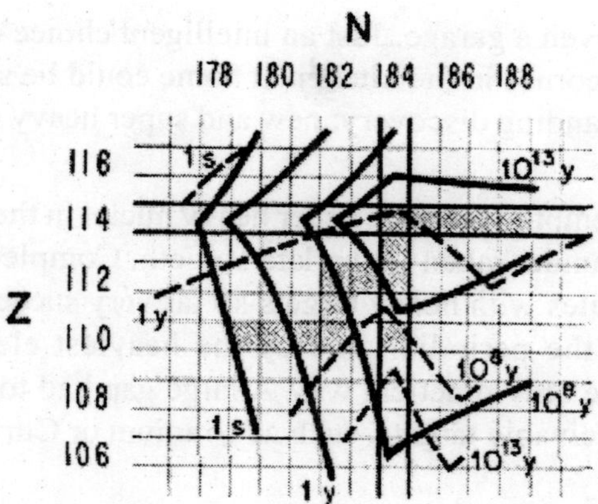

Fig. 2. Topology of the Island of Super Heavy Nuclei Around the Shell Closures at Proton Number Z=114 and Neutron Number N=184.

In the above figure, thick solid lines are contours of spontaneous fission half-lives and broken lines refer to á decay half lives. Considering the three major decay modes, spontaneous fission, alpha (á) decay and beta (â) decay or electron capture, spontaneous fission half-lives were calculated to peak sharply at the doubly magic nucleus, descending by orders of magnitude with in short distances in the Z-N plane, thus causing the island like shape. In contrast, half lives for alpha-decay should decrease rather uniformly with increasing proton number with some zigzag at the nuclear shell closures. Beta stable nuclei would cross the plane as a diagonal belt. For Z=114, N=184 an enormous spontaneous fission half life of $2x10^{19}$ yrs was estimated, sufficiently long for the occurrence of super heavy elements in nature. Additional stability was expected for odd elements such as 111 or 113 due to the well known hindrance of spontaneous fission and á decay for odd proton numbers. Such predictions of very long over all half lives immediately stirred up a gold rush period of hunting for super heavy elements in natural samples. Everybody could feel encouraged to participate. Nearly nothing would be needed; little money, no equipment, no research group, no permissions by the laboratory director, no accelerator beam time, no proposal to finding agencies not even a garage. Just an intelligent choice of a natural sample and a corner in the kitchen at home could be sufficient to make an outstanding discovery; new and super heavy elements in nature.

First attempt to produce super heavy nuclei in the laboratory were already under taken in the late sixtieth. Complete fusion of heavy projectiles with heavy targets-so far very successful in the extension of the periodic table by the heaviest elements was considered the only practical way. A large gap had to be bridged between conceivable targets, such as Uranium or Curium and the islands.

The crucial question was, where are the SHE located in the periodic table and how well do they fit its architecture? The answer had immediate implications for the ongoing "search for" campaigns either for the selection of natural samples or the design of chemical identification procedures in synthesis experiments.

In a naive continuation of the table, element 110 is located below Pletinum, element 112 below Mercury, element 114 below Lead and 118 becomes the next noble gas below radon. Quantum Mechanical calculations of ground state electronic configuration supported this idea..

Existence of Super Heavy Elements

The existence of valley in the nuclear chart (Fig. 3) is explained by the separation energy of nucleons which is defined as the energy required to remove one nucleon (proton or neutron) from the nucleus.

The separation energy of a neutron is –

$$S_n(N,Z) = B_e(N,Z) - B_e(N-1,Z)$$

Similarly, for the proton, the separation energies-

$$S_p(N,Z) = B_e(N,Z) - B_e(N-1,Z)$$

By adding neutrons or protons to a nucleus, the nucleus does not accept any more nucleons after a certain number, so at this limit the separation energy of the proton or neutron has reached zero.

In the valley of stability, the heaviest stable nucleus is $_{82}Pb^{208}$ with (N=126, Z=82) which is a doubly magic nucleus. According to nuclear shell model, the next doubly magic stable nucleus heavier than Pb is predicted to be at (Z=104, N=184) with in sea of instability. This island of stability is so called the region of super heavy elements.

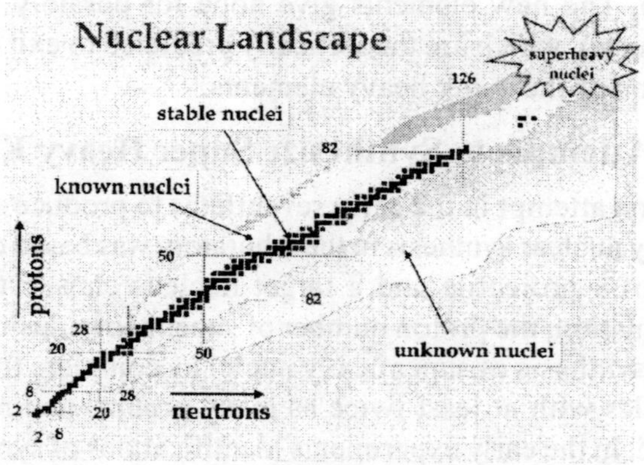

Fig. 3. Nuclear Chart

Search for Super Heavy Elements in Nature

The subject of searching for very long lived nuclei, and neutron rich nuclei was a highly extensive issue during 1970 and 1980. The main search was to find the possibility of heavier elements than those naturally occurring such as uranium that can exit in the natural samples such as soil, rocks and in the cosmic rays.

The searching arguments were passed on the controversy of the different models that predicted the decay properties of these nuclei. One of these decay modes is the spontaneous fission which is predominant in the region of heavy nuclei to be found in the earth's crust or in nature in general.

Most of the experiments were built for the purpose of investigation the heavier elements in same object samples. The resultant of these investigation led to the detection of the elements Hassium (Z=108). Searching for the heaviest nuclei possible is an interesting and open challenging issue in nuclear structure. One important question for instance is to identify the maximum number possible of nucleons to be bound into a single nucleus. By using neutron rich beam with a neutron rich target, scientists were able to discover some of the heavier nucleus, perhaps even heavier then uranium. The new elements give hope for possibly lived super heavy nuclei, which are theoretically predicted to exit and to form the region of the super heavy elements.

Early Attempts to Synthesize Super Heavy Elements

First attempt in the early seventieths to produce super heavy nuclei by nuclear synthesis in the laboratory was based on complete fusion of a projectile and a target nucleus chosen to attain by amalgamation, the proton number of super heavy desired element. For the synthesis of super heavy nuclei by complete fusion, larger projectiles with at least twice as many protons and neutrons are required. In the early experiments identifications of synthetic super heavy nuclei was mainly attempted by radioactivity detection.

Synthesis of She

Super heavy elements have all been synthesized during the latter half of the twentieth century and they are still being produced during the twenty first century as high advanced technology. These elements are produced by the bombardment of lighter elements using particle accelerators namely using fusion reaction to create them.

For ex. $${}^{70}_{30}Zn + {}^{208}_{80}Pb \rightarrow {}^{277}_{112}Uub + n$$

Synthesizing heavy elements by fusion reactions requires choosing good methods of projectile and target that yield nuclei with less deformation and a particularly stable nucleus. The most difficult issue in the experiments is how to choose a reaction that does not yield to an excited nucleus which might decay by fission instead of creating a heavier and stable nucleus. For this reason, scientists are switching from using light projectiles with heavy targets to less massive targets with relatively heavier projectile. Ex- Pb-208 or Bi-209 with Cr-54 or iron-58.

Fusion Reaction

Nuclear fusion is the process in which multi atomic nuclei join together to form a heavier nucleus. It is followed by the release or absorption of energy. Iron and nickel nuclei are the most stable nuclei and they have the largest binding energy per nucleon. Therefore a fusion reaction of nuclei lighter than iron and nickel release energy and reactions of nuclei heavier than iron and nickel absorbs energy. Nuclear fusion occurs naturally in stars and also has been made artificially but it is not completely controlled. Fusion reactions are of two types as discussed below:

Cold Fusion

In cold fusion reactions, the excitation energy of the fused nuclei remains low. After the reactions, it cools down the nucleus to its ground state by emission of one or two neutrons or gamma rays. This the main advantage of this type of reactions.

In the reactions of the fusion of lead-208 or bismath-208 magic nuclei with heavy projectiles, the maximum result of the reaction products with atomic number (A=104-112) is observed experimentally at a compound nucleus with energy (E=20-11Mev) respectively. The reaction products are typically created via successive alpha-decays ending by spontaneous fission (SF) in the region of the known elements. Their identification is facilitated through "alpha-alpha", or parent-daughter correlations. Cold fusion reactions have been used at GSI to produce Elements 107 to 112. A reasonable estimate of the limit of cold fusion experiments comes from the measurement of a single isotope of Element 113 which was synthesised at RIKEN, Japan with the extremely small production cross-section on the order of Femto-barns. Synthesis with ^{208}Pb target using cold fusion reaction is shown in Fig. 4.

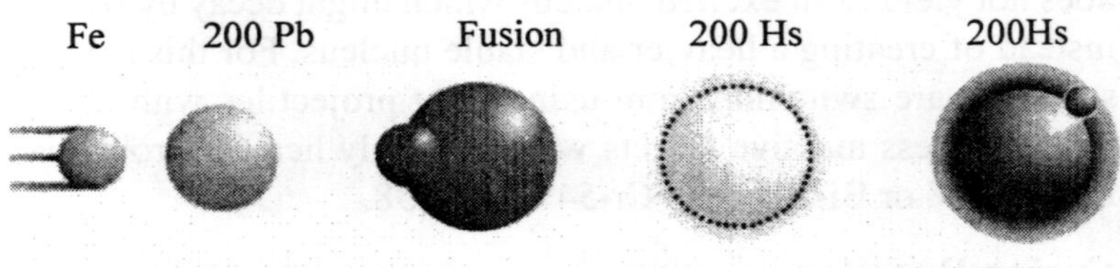

Fig. 4. Synthesis of SHE by Cold Fusion

Hot Fusion

Using hot fusion reaction to produce super heavy elements (Fig. 5) always involves a heavy target in the region of actinide nuclei with light projectile. The heavy target includes many protons which give rise to the coulomb barrier. In order to make very heavy nuclei by a fusion reaction, we need to cool down the hot compound nucleus which has more excitation energy. The life time of these nuclei is limited by the possibility of the fission into two parts, so that super heavy nucleus survives. It is by evaporation of neutron from the compound nucleus.

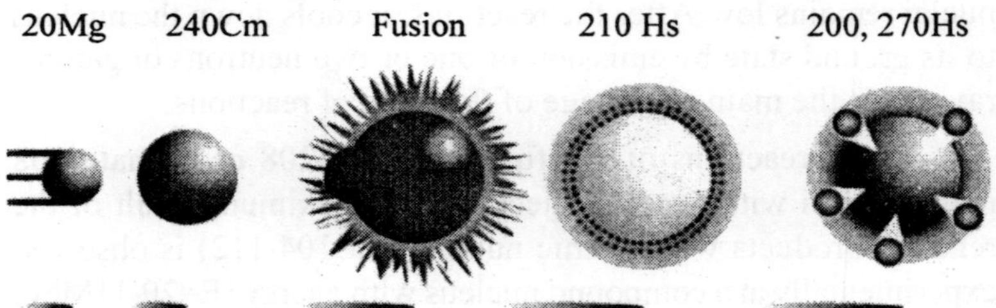

Fig. 5. Synthesis of SHE by Hot Fusion

Hot fusion has been used successfully to produce the more neutron rich species ranging from Rutherfordium to Element 118. Unlike cold fusion, these decay chains end in unknown regions presenting additional challenges to the conclusive identification of the nuclides especially in the absence of elemental or isotopic signatures. Using hot fusion reaction, five super heavy elements have been produced successfully namely, Nobelium (No, Z=102), Lawrencium (Lr, Z=103), Rutherfordium (Rf, Z=104), Dubnium

(Db, Z=105) and Seaborgium (Sg, Z=106) and elements 113 to 118 have been produced with hot fusion at FLNR using actinide targets from uranium to californium with beams of the extremely rare isotope - ^{48}Ca.

Elements Made at Jinr and GSI

Below we list some elements which are produced at Joint Institute for Nuclear Research in Dubna, Russia (then the Soviet Union) and 'Gesellschaft für Schwerionenforschung, GSI (Society for heavy ion research) in Darmstadt, Germany.

Table 1. Elements prepared at JINR and GSI

Atomic No.	Name	Symbol	Named after	Institute	Generation
104	Rutherfordium	Rf	Ernest Rutherford (physicist)	JINR	1966
105	Dubnium	Db	City of Dubna, Russia	JINR	1968
106	Seaorgium	Sg	Glenn T. Seaborg (physicist)	JINR	1974
107	Bohrium	Bh	Niels Bohr (Danish physicist)	JINR	1981
108	Hassium	Hs	Hessen (the German Bundesland)	GSI	1984
109	Meitnerium	Mt	Lise Meitner (Austrian physicist)	GSI	1982
110	Darmstadtium	Ds	Darmstadt (city in Germany)	GSI	1994
111	Roentgenium	Rg	Wilhelm Conrad Rontgen (discoverer of X rays)	GSI	1994
112	Copernicium	Cn	Nicolaus Copernicus (Astronomer)	GSI	1996

Detection and Identification of she

There are several techniques used to separate the super heavy elements nuclei form the products of the reaction after synthesizing them. The techniques such as recoil separators, are designed to filter and separate the produced nuclei from fusion reactions with a high rate of transmission (Fig. 6). The transmitted particles have to be detected by a detector System located within the synthesis system and the choice of such a suitable detector depends on the energy and rate of the particles ,their half lives and their type of decay mode of the particles.

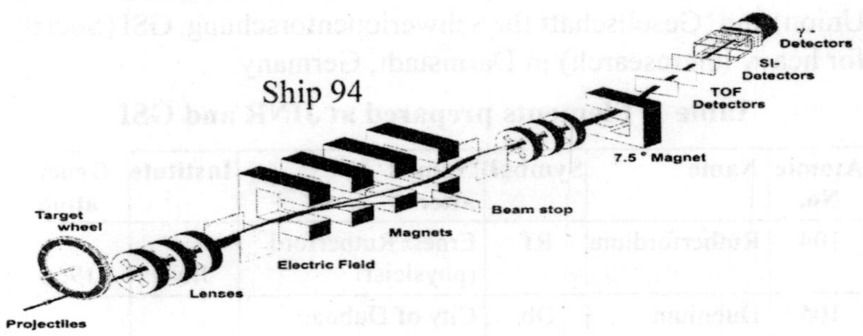

Fig. 6. Separator for Heavy Ion Reaction Products (SHIP)

Experimental and theoretical evidences show that solid state detectors (semi conductor detectors) are the appropriate detectors to detect and identify super heavy nuclei and also to measure their decay properties, semi - conductor detectors are mostly designed by using high purely Germanium or Silicon and sometimes by using junction of same elements. The.decay signal from the super heavy nuclei transmits in to the effective volume of the semi conductor detectors. The particles transfer their energy to the valence band electrons which enable them to move toward the conduction band. This flow of the electrons can be measured as a current which gives an indication for energy of the incident particles.

Recent Discoveries

In different laboratories such as JINR (Joint institute of nuclear research) in Dubna , GSI in Germany and LBNL in USA , where the reaction of synthesizing new elements took place, experimental attempts are still running in a competitive way to

Table 2. List of observed super heavy elements

Atomic No.	Atomic mass	Expected half-life (sec.)
118	294	0.4×10^{-3}
116	293	80×10^{-3}
	292	40×10^{-3}
	291	20×10^{-3}
	290	10×10^{-3}
115	288	60×10^{-9}
	287	30×10^{-9}
114	289	2.0
	288	0.9
	287	0.5
	286	0.2
113	284	0.3
	283	140×10^{-3}
	282	6×10^{-3}
112	284	
	283	3.0
	282	0.4
111	280	7×10^{-3}
	279	1×10^{-3}
	278	
110	281	
	279	0.2
109	276	0.1
	275	2×10^{-3}
	274	20×10^{-3}
108	275	0.8
107	272	3.0
	270	5.0
106	271	48
	105	268
	266	
104	267	

answer the question of what is the limit of creating super heavy element. The heaviest element deemed discovered to date is element 112 by the IUPAC, GSI being the laboratory credited with its discovery, proposed the name coperricium with chemical symbol 'cn'.

A collaboration of Russian and US physicists has created the SHE 117, an experimental achievement that fills in the final gap on the list of observed elements.

Below we are giving a table which comprises of recently observed super heavy elements.

Future Aspects

The progress towards the exploration of the island of spherical SHEs is difficult to predict. However, despite the exciting new results, many questions of more general character are still unanswered. New developments will not only make it possible to perform experiments aimed at synthesizing new elements in reasonable measuring times, but will also allow for a number of various other investigations covering reaction physics and spectroscopy.

One can hope that, during the coming years more data will be measured. In order to promote a better understanding of the stability of the heaviest elements that lead to fusion. A microscopic description of the fusion process will be needed to an effective explanation of all measured phenomena in the case of low dissipative energies. Then, the relationships between fusion probability and stability of the fusion products may also become apparent.

References

1. **Yuri. Oganessian.** Study of Heavy Nuclei at Flnr(dubna). *European Physical Journal* D, 45:17–23, 2007.
2. **B.W. Petley P.J.Karol, H. Nakahara and E. Vogt.** On the Claims for Discovery of Elements 110,111,112,114,116, and 118. *Pure and Applied Chemistry*, 75:1601–1611, 2003.

3. **G.T. Seaborg and W. Loveland.** *Superheavy Elements*. Contemp. Phys., 28:33–48, 1987.
4. wikipedia. nuclear fusion.
5. James Wittke. www4.nau.edu.
6. **Yu.Ts.Oganessian.** Reactions of Synthesis of Heavy Nuclei:Breif Summery and Outlook. *Physics of atomic nuclei*, 69:932–940, 2006.
7. **Y. T. Oganessian *et al.*,** *Phys. Rev. Lett.* 104, 142502 (2010).
8. **P. Möller *et al.*,** *Phys. Rev.* C 79, 064304 (2009).
9. **S. Hofmann and G. Münzenberg,** *Rev. Mod. Phys.* 72, 733 (2000).
10. **K. Morita *et al.*,** *J. Phys. Soc. Jpn.* 73, 2593 (2004).
11. **Yu. Ts. Oganessian,** *J. Phys.* G 34, R165 (2007).
12. **V. Zagrebaev and W. Greiner,** *Phys. Rev.* C 78, 034610 (2008).

19

MOSSBAUER EFFECT

K. Chaturvedi, Anuj Mishra and Ashish Nautiyal

Introduction

Mossbauer was the first to observe this effect in 1958. He used gamma rays of energy 0.129 MeV, corresponding to the transition from ground to the first excited state $\frac{191}{77}Ir$. He was awarded the Nobel Prize in 1961. The Mossbauer Effect has also been used to verify the prediction of gravitational red shift.

Mossbauer Effect is often used in Mossbauer spectroscopy. In the emission of γ-ray photons from an atom corresponding to a transition to nuclear ground state, momentum must be conserved, so the atom must have small recoil. Energy balance then implies that gamma rays are emitted with a spread of energies. When an atom is part of a crystal lattice, however the entire lattice may recoil resulting in a quantized vibrational energy termed a phonon. If no phonon is emitted or absorbed, the emitted gamma rays have a very small spread of energies, given by

$$\Delta E = \frac{h}{T} .$$

Description of the Experiment

In general, gamma rays are produced by nuclear transitions from an unstable high-energy state, to a stable low-energy state. The energy of the emitted gamma ray corresponds to the energy of the nuclear transition, minus an amount of energy that is lost as recoil to the emitting atom. If the lost "recoil energy" is small compared with the energy line width of the nuclear transition, then the gamma ray energy still corresponds to the energy of the nuclear transition, and the gamma ray can be absorbed by a second atom of the same type as the first. This emission and subsequent absorption is called resonance. Additional recoil energy is also lost during absorption, so in order for resonance to occur the recoil energy must actually be less than half the line width for the corresponding nuclear transition.

The amount of energy in the recoiling body (E_R) can be found from momentum conservation:

$$P_R| = |P_\gamma|$$

$$P_R|^2 = |P_\gamma|^2$$

Where P_R is the momentum of the recoiling matter and P_γ the momentum of the gamma ray. Substituting energy into the equation gives:

$$2ME_R = \frac{E_\gamma^2}{c^2}$$

$$E_R = \frac{E_\gamma^2}{2Mc^2}$$

Where E_R (=0.002 eV for ^{57}Fe) is the energy lost as recoil, E_γ is the energy of the gamma ray (=14.4 keV for ^{57}Fe), M (=56.9354 u for ^{57}Fe) is the mass of the emitting or absorbing body, and c is the speed of light.

In his original experiment, Mossbauer measured the emission of 129 keV gamma ray from radioactive β^- source Osmium (^{191}Os). They were passed through a metallic iridium absorber (39% of ^{191}Ir) and then on to a detector. Gamma source ^{191}Ir was formed in an isomeric state in the beta decay of ^{191}Os. The natural width and life time of the 129 keV levels of ^{191}Ir are 5×10^{-6} eV and 1.3×10^{-10} seconds respectively.

The recoil energy of the free nucleus for 129 keV gamma radiations would be 0.047 eV and the Doppler broadening at room temperature about 0.1 eV. Mossbauer measured the effective absorption cross section per nucleus of ^{191}Ir for radiation of 129 keV as a function of temperature by cooling the source and absorber with liquid air from about 400°K to 88°K (Fig. 1). He found that the absorption cross-section increased sharply as the temperature was decreased.

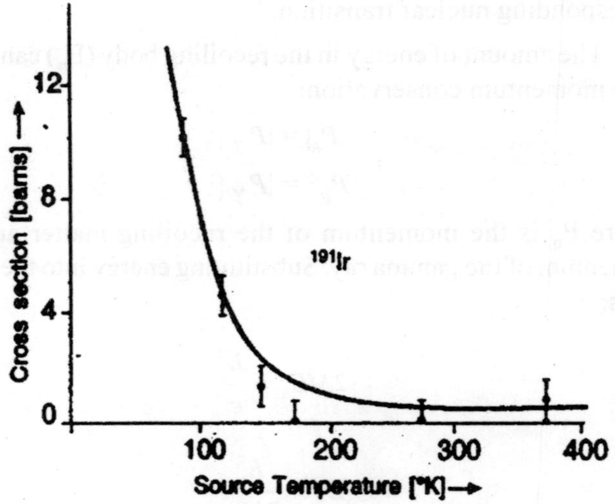

Fig. 1: The Variation of Absorption Cross Section with Temperature in ^{191}Ir for 129 keV γ-rays

In another experiment Mossbauer kept both the source and the absorber at 88°K and imposed a very low velocity on the source. It was done by mounting the source on a turn table and rotating it very slowly. The systematic diagram of the experimental arrangement is shown in Fig. 2. Here A is the absorber in a cryostat and S is the source at the rim of rotating cryostat and M is the region in which the source is seen from the scintillation detector. If one moves the source with respect to the absorber the frequency of the absorption line changes by an amount.

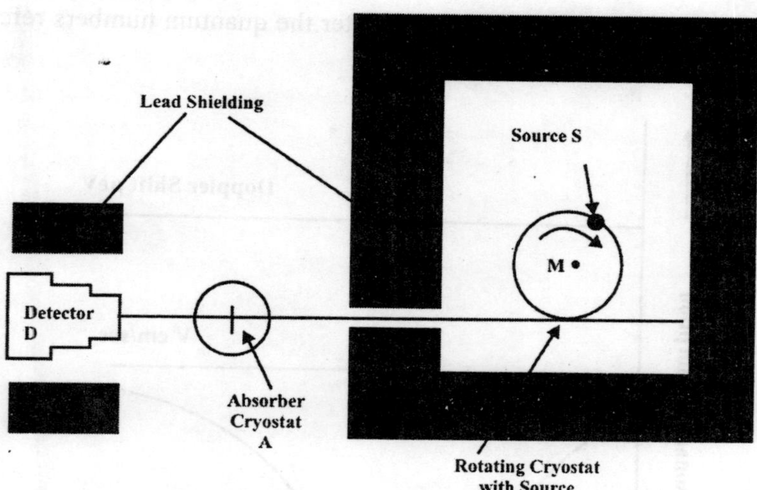

Fig. 2: Experimental Arrangement by Mossbauer

$$\Delta\omega / \omega = \Delta v / c \, or \Delta E = (\Delta v / c)E_0,$$

where, Δv is the relative velocity and ΔE is the Doppler shift

The fluorescence absorption in ^{191}Ir as a function of relative velocity v between the source and the absorber is shown in Fig 3. The percentage absorption is a maximum at zero relative velocity. The curve so obtained has a width at half maximum of 2 cm/sec corresponding to an energy width of 9.2 ± 1.2 keV. Interpreting this value as twice the natural width $\Gamma = 4.6 \pm 0.6$ keV, Mossbauer obtained the value $\tau = (1.4^{+0.2}_{-0.1}) \times 10^{-10}$ sec. for the lifetime of the 129 keV state in ^{191}Ir.

Physicists not only repeated these experiments but investigated other nuclei, suitable for Mossbauer Effect. The recoilless emission of an extremely narrow 14.4 keV ã-line from the first level of ^{57}Fe was observed at several centres simultaneously. ^{57}Fe is produced by the process of orbital electron capture in ^{57}Co, which has a half-life of 270 days and is generally prepared by deuteron bombardment of an iron target in the form of a strip. Decay schemes of osmium and cobalt are given in Fig 4.

The positive and negative signs after the quantum numbers refer to the parity of the states.

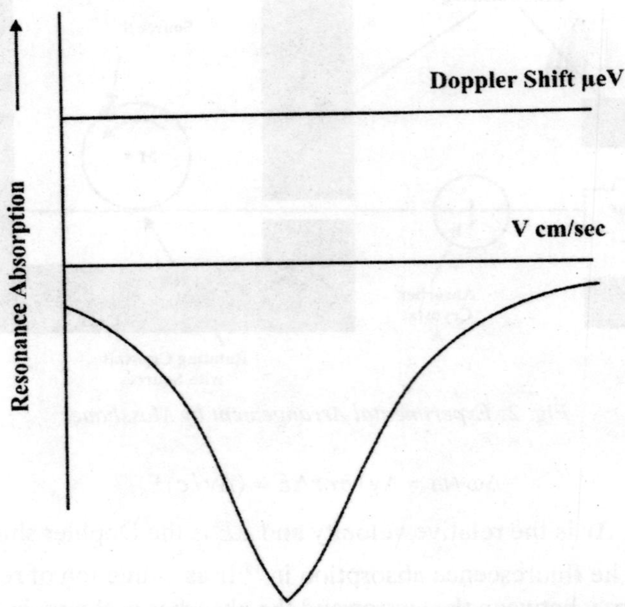

Fig. 3: Resonance Absorption in ^{191}Ir as a Function of Relative Velocity Between Source and Absorber

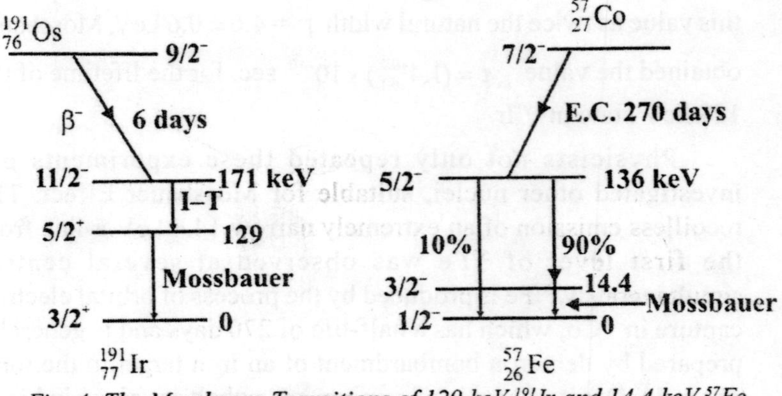

Fig. 4: The Mossbauer Transitions of 129 keV ^{191}Ir and 14.4 keV ^{57}Fe.

The 14.4 keV level in ^{57}Fe has a mean life of 1.4×10^{-7} sec and the corresponding line width is 4.6×10^{-7} eV. This life time is long enough for the excited ^{57}Fe ions to occupy suitable sites in the iron crystal lattice before the decay. The 14.4 keV gamma rays can be passed through an iron absorber which can be enriched in ^{57}Fe so as to increase the probability of recoilless absorption. The gamma rays can be detected by a proportional counter or scintillation counter.

After ^{57}Fe, the substance employed most frequently is probably tin-119 and a fair amount of work has also been done with same rare earth elements. The theory of recoilless emission shows that the effect is greater when: (*a*) the gamma ray energy E is small (The best example is the 14.4 keV transition of ^{57}Fe); (*b*) the temperature of the crystal source is small; (*c*) the Debye temperature of the crystal lattice is high.

The theoretical analysis, following Lamb's treatment, used Debye continuum theory to describe lattice vibration behavior in terms of a model based on a system of independent linear oscillators with a continuous frequency distribution upto a maximum ω_D. It is related with Debye temp (H), as

$$h_{\omega D} = k(H),$$

where, k is the Boltzmann constant. The fraction of recoilless γ-transition, also known as Lamb Mossbauer factor is given as

$$f \cong \exp\left[-\frac{3E_R}{2K(H)}\left\{1+\frac{2}{3}\left(\frac{\pi T}{H}\right)^2\right\}\right]$$

The first term is independent of temperature. It shows that even at T = 0. f is large only if the recoil energy of the free nucleus is small compared to k(H). The probability of recoilless decay decreases with

In order to utilize the Mossbauer effect, one has to consider transitions in which

- f is adequately large;

- E_0 must be fairly large for a good precision;
- There must be small livelihood of internal conversion;
- Fine structure splitting should not occur, if possible.

Applications of Mossbauer Effect

Isomer Shift or Chemical Shift-Different chemical states of an atom are associated with differences in the electron distribution around the nucleus and contribute to the energy of transition.

$$E_\gamma = \Delta E_{nuc} + \Delta E_{elec}$$

where, ΔE_{nuc} is the change in nuclear binding energy and ΔE_{elec}, is the change in B.E. of the atomic electrons. If the emitting nucleus (excited state) and the absorbing nucleus (ground state) are in different chemical compounds, the distributions in atomic electrons will be different, which will cause differences in ΔE_{elec}, and thus in E_γ. The change in E_γ is called the isomer shift or chemical shift (as it is related to the chemical environment of the atom). The Mossbauer technique is used to show the energy shift due to the minute effects of chemical binding on the nuclear level structure.

Magnetic Hyperfine Splitting

If either the emitting or the absorbing nucleus has a spin $I \geq \frac{1}{2}$, it will also have a magnetic moment. In the presence of magnetic field, the energy of the nucleus will depend on its orientation with respect to the magnetic field. The projection of spin I may take 2I+1 values. The absorption spectrum obtained by the Mossbauer technique will thus show a hyperfine splitting into 2I + 1 parts. In most measurements of the nuclear hyperfine interactions the line splitting observed resulted from the interaction of the nuclear moments with internal fields. These fields are generally too weak to produce splitting larger than the natural widths of the γ-Iines $(2\Gamma_0)$.

Since iron is ferromagnetic, the magnetic field intensity at the nucleus of iron-57 produced by the external electrons is large enough to produce a splitting of the energy levels, depending upon the value of I. This splitting is the nuclear equivalent of a Zeeman effect. The energy separation of these levels will be given by $\Delta E = Bg\mu_N$ where μ_N is the nuclear magneton. The number of counts of components establishes the spin I and the separation M at a known field B evaluates the Lande factor g. Thus the magnetic moment μ can be calculated from g and I.

In the case of ^{57}Fe, the value of I for the ground state is $1/2^-$ and for the first excited state is $3/2^-$. The selection rule $\Delta m_1 = 0, \pm 1$, , predicts the splitting of the γ-ray line into six components, as shown in Fig. 5.

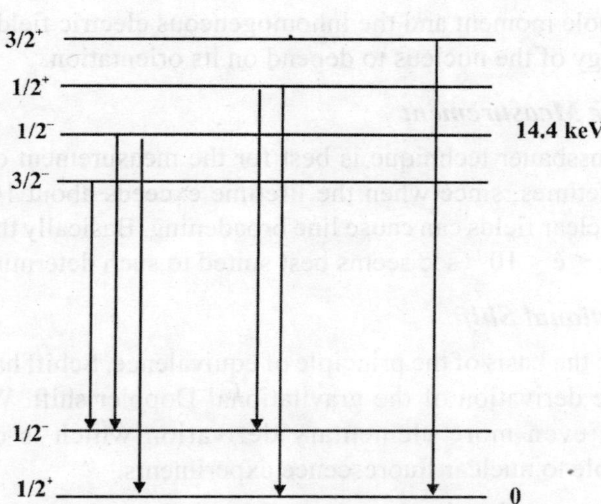

Fig. 5. Nuclear Hyperfine Structure of 14.4 keV Transition in ^{57}Fe

If the source is moving relative to the absorber, the change in energy of the radiation due to Doppler effect will occur [MD = (v/c)E]. If both the source and absorber are ferromagnetic; there will be some velocities for which the energy change will be equal to the separations of some of the energy levels, so that the several components may overlap. Thus a non-ferromagnetic material, such as stainless steel, should be used as either the source or the absorber.

The nuclear hyperfine structure of ^{57}Fe in iron metal is obtained with a stainless steel source and a natural iron absorber 0.001 inch thick, as observed by G.K.

In the earlier experiment by Hanna et al. (1960), an external magnetic field was applied to both the source and the absorber and observed the nuclear hyperfine lines for ^{57}Fe. The effective value of the internal magnetic field at the ^{57}Fe nucleus was established to be $(3.33 \pm 0.10) \times 10^5$ oersteds.

Quadrupole Hyperfine Splitting

If either the emitting or the absorbing nucleus has a spin $I \geq 1$ and is in an inhomogeneous electric field, energy may split in several lines. It is because the interaction between the nuclear quadrupole moment and the inhomogeneous electric field causes the energy of the nucleus to depend on its orientation.

Lifetime Measurement

Mossbauer technique is best for the measurement of fairly short lifetimes, since when the lifetime exceeds about 10^{-10} sec, extra nuclear fields can cause line broadening. Basically the range 10^{-13} sec $\leq$ é $\leq 10^{-10}$ sec seems best suited to such determinations.

Gravitational Shift

On the basis of the principle of equivalence, Schiff has given a simple derivation of the gravitational Doppler shift. We shall give an even more elementary derivation which is directly applicable to nuclear fluorescence experiments.

We know that the intensity I of the earth gravitational field at a distance R from an earth of mass M is the acceleration due to gravity. Hence

$$I = GM/R^2g$$

Similarly the gravitational potential ö at the point R due to earth is

$$\varphi = - GM/R = - gR.$$

Let us place a source nucleus A and an observer nucleus B as shown in Fig. 6.

Fig. 6. Table PQ has Constant Acceleration f in Inertial Frame of Reference

Table PQ has constant acceleration f when the velocity of table is v_1. The nucleus A at one end emits a gamma ray photon of frequency ν_l in the direction of its motion. The frequency of this radiation as measured by a stationary observer will be

$$\nu_{ob} = \nu_l (1+ v_1/c).$$

This photon moves towards B with a constant velocity c. It will reach an identical absorber nucleus B at a time $t(= l/c)$. The velocity of table at this time is $v_2 = v_1 + ft$. The apparent frequency of radiation, which can be absorbed by observer nucleus B will be ν_2. The frequency of this radiation as measured by a stationary observer will be

$$\nu_{ob} = \nu_2 (1+v_2/c)$$

Clearly if the gamma ray absorbed by nucleus B is to be that one which has been emitted by nucleus A, we must have

$$\nu_l (1+v_1/c) = \nu_2(1+v_2/c)$$

This relation indicates that absorption would take place for identical recoilless nuclei, only if $\nu_1 = \nu_2$. If we assume that $\nu_1 - \nu_2 = \Delta\nu$ and $\nu_1 = \nu_2 = \nu$, then we have

$$\Delta\nu = \frac{\nu}{c}(v_2 - v_1) = \frac{\nu}{c} f \frac{l}{c} or \frac{\Delta\nu}{\nu} = \frac{fl}{c^2}$$

The principle of equivalence of an accelerated system tells us that the system having an acceleration f produces an effect equivalent to that of a gravitational acceleration – f (= g). If $\Delta\phi$ is the potential difference between two points in a gravitational field, the difference in frequency of identical clocks at these points will be

$$\Delta\nu = \nu\Delta^{\phi\frac{\square}{c^2}}$$

Thus for resonance absorption, the frequency of emission v_1 must be larger than the frequency of absorption v_2 by an amount $v\Delta\phi/c^2$. Hence the frequency of radiation coming from a source located below the absorber will be shifted towards the red.

Until 1960, the gravitational red shift could only be observed with light coming from stars, as $\Delta\phi$ is very large in that case. Apart of above applications, Mossbauer effect can be used in Mossbauer spectroscopy. It can be used to verify the prediction of gravitational red shift. It also provides information about the chemical environment which can be used to characterize the sample. Mossbauer effect can be used for semi – quantitative analysis and in the nuclear physics, we use Mossbauer effect for the study of absorption or emission of the energy level transition.

Mossbauer Spectroscopy

The gamma ray source is a radioactive element that is mechanically vibrated back and forth to Doppler shift the energy of the emitted gamma radiation. The schematic below (Fig. 7) shows a transmission Mossbauer experiment. As the energy of the gamma radiation is scanned by Doppler shifting, the detector records the frequencies of gamma radiation that are absorbed by the sample.

What Mossbauer discovered is that when the atoms are within a solid matrix, the effective mass of the nucleus is very much greater. The recoiling mass is now effectively the mass of the whole system, making ER and ED very small. If the gamma-ray energy is small enough the recoil of the nucleus is too low to be transmitted as a phonon (vibration in the crystal lattice) and so the whole system recoils, making the recoil energy practically zero, a recoil-free

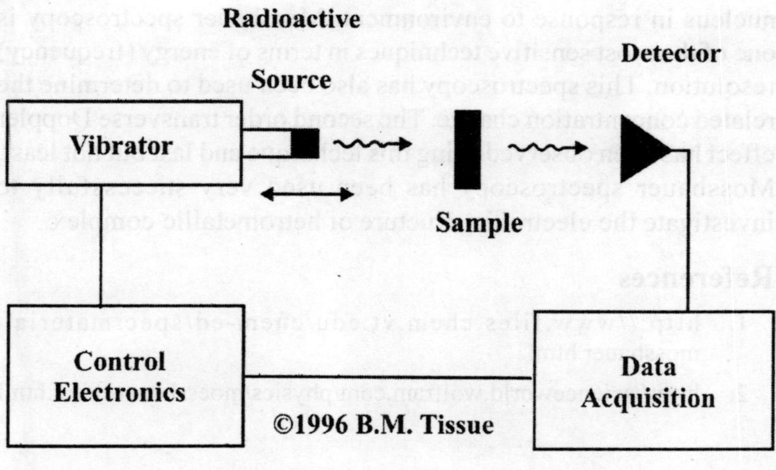

Fig. 7. Schematic of an Experimental Set-up for Mossbauer Spectroscopy

event. In this situation, as shown in Fig 8, if the emitting and absorbing nuclei are in a solid matrix, the emitted and absorbed gamma ray is of the same energy.

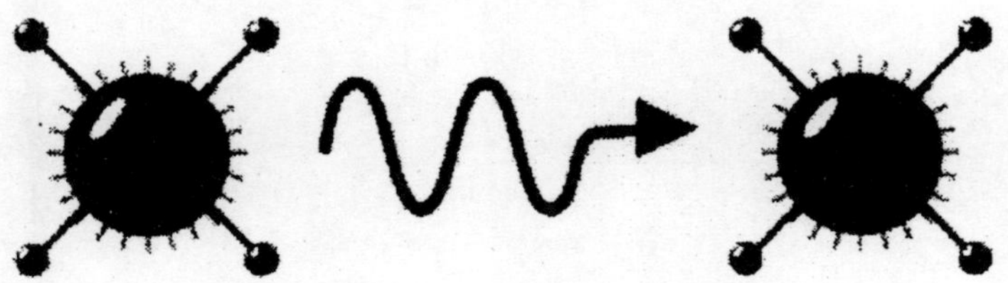

Fig. 8. Recoil-free Emission or Absorption of a Gamma-ray When the Nuclei are in a Solid Matrix Such as a Crystal Lattice

Uses of Mossbauer Spectroscopy

Mossbauer spectroscopy is a versatile technique that can be used to provide information in many areas of science such as physic, chemistry & Biology. This probes the energy levels of an atomic nucleus in response to environment. Mossbauer spectroscopy is one of the most sensitive techniques in terms of energy (frequency) resolution. This spectroscopy has also been used to determine the related concentration change. The second order transverse Doppler effect has been observed using this technique and last but not least, Mossbauer spectroscopy has been used very successfully to investigate the electronic structure of hetrometallic complex.

References

1. http://www.files.chem.vt.edu/chem-ed/spec/material/mossbauer.html
2. http://scienceworld.wolfram.com/physics/moessbauereffect.html

20

RADIATION HAZARDS: CAUSES & EFFECTS

K. Chaturvedi, Vidya and Rahul

Introduction

In physics, *radiation* is a process in which energetic particles or energy or waves travel through a medium or space. There are two distinct types of radiation; ionizing and non-ionizing. Both ionizing and non-ionizing radiation can be harmful to organisms and can result in changes to the natural environment.

(A) Ionizing radiations are of following types:

- Alpha
- Beta
- Neutron
- Gamma and X-ray

(B) Non – ionizing radiations are

- Neutron radiation
- Electromagnetic radiation
  - visible light
  - Infrared
  - Microwave
  - Radio waves

- Very low frequency (VLF)
- Extremely low frequency (ELF)
- Thermal radiation (heat)
- Black body radiation

Ionizing Radiation

Radiation with sufficiently high energy can ionize atoms. Because cells and specially the DNA can be damaged, this ionization can result in an increased chance of cancer. An individual cell is made of trillions of atoms. The probability of ionizing radiation causing cancer is dependent upon the dose rate of the radiation and the sensitivity of the organism being irradiated. Alpha particles, beta particles, gamma rays, X-ray radiation, and neutrons may all be accelerated to energy high enough to ionize atoms.

Alpha Radiation

Alpha radiation is a heavy, very short-range particle and is actually an ejected helium nucleus. Some characteristics of alpha radiation are:

- Most alpha radiation is not able to penetrate human skin.
- Alpha-emitting materials can be harmful to humans if the materials are inhaled, swallowed, or absorbed through open wounds.
- A variety of instruments has been designed to measure alpha radiation. Special training in the use of these instruments is essential for making accurate measurements.
- A thin-window Geiger-Mueller (GM) probe can detect the presence of alpha radiation.
- Instruments cannot detect alpha radiation through even a thin layer of water, dust, paper, or other material, because alpha radiation is not penetrating.

Examples of some alpha emitters: radium, radon, uranium, thorium.

Beta Radiation

Beta radiation is a light, short-range particle and is actually an ejected electron. Some characteristics of beta radiation are:

- Beta radiation may travel several feet in air and is moderately penetrating.
- Beta radiation can penetrate human skin to the "germinal layer," where new skin cells are produced. If high levels of beta-emitting contaminants are allowed to remain on the skin for a prolonged period of time, they may cause skin injury.
- Beta-emitting contaminants may be harmful if deposited internally.
- Most beta emitters can be detected with a survey instrument and a thin-window GM probe (e.g., "pancake" type). Some beta emitters, however, produce very low-energy, poorly penetrating radiation that may be difficult or impossible to detect.

Examples of these difficult-to-detect beta emitters are hydrogen-3 (tritium), carbon-14, and sulfur-35.

Examples of some pure beta emitters: strontium-90, carbon-14, tritium, and sulfur-35.

Neutron

Neutrons are categorized according to their speed. Some characteristics are:

- High-energy (high-speed) neutrons have the ability to ionize atoms and are able to deeply penetrate materials.
- Neutrons are the only type of ionizing radiation that can make other objects, or material, radioactive. This process, called neutron activation, is the primary method used to produce radioactive sources for use in medical, academic, and industrial applications.
- High-energy neutrons can travel great distances in air and typically require hydrogen rich shielding, such as concrete or water, to block them.

A common source of neutron radiation occurs inside a nuclear reactor, where many feet of water is used as effective shielding.

Gamma Radiation and X Rays

Gamma radiation and X-rays are highly penetrating electromagnetic radiation. Some characteristics of these radiations are:

- Gamma radiation or x rays are able to travel many feet in air and many inches in human tissue. They readily penetrate most materials and are sometimes called "penetrating" radiation.
- X-rays are like gamma rays. X-rays, too, are penetrating radiation. Sealed radioactive sources and machines that emit gamma radiation and X-rays respectively constitute mainly an external hazard to humans.

Gamma radiation and X-rays are electromagnetic radiation like visible light, radiowaves, and ultraviolet light. These electromagnetic radiations differ only in the amount of energy they have. Gamma rays and X-rays are the most energetic of these.

- Gamma radiation is easily detected by survey meters with a sodium iodide detector probe.
- Gamma radiation and/or characteristic X-rays frequently accompany the emission of alpha and beta radiation during radioactive decay.

Examples of some gamma emitters: iodine-131, cesium-137, cobalt-60, radium-226, and technetium-99m.

The relative abilities of three different types of ionising radiations (α, β & γ) to penetrate solid matter are illustrated in Fig. 1. Alfa particles (α) are stopped by a sheet of paper while beta particles (β) are stopped by an aluminium plate. Gamma radiation (γ) is dampened when it penetrates matter.

Non-Ionizing Radiation

The energy of non-ionizing radiation is less and instead of producing charged ions when passing through matter, the electromagnetic radiation has only sufficient energy to change the rotational, vibrational or electronic valence configurations of molecules and atoms. The effect of non-ionizing forms of radiation on living tissue has only recently been studied. Nevertheless, different biological effects are observed for different types of non-ionizing radiation.

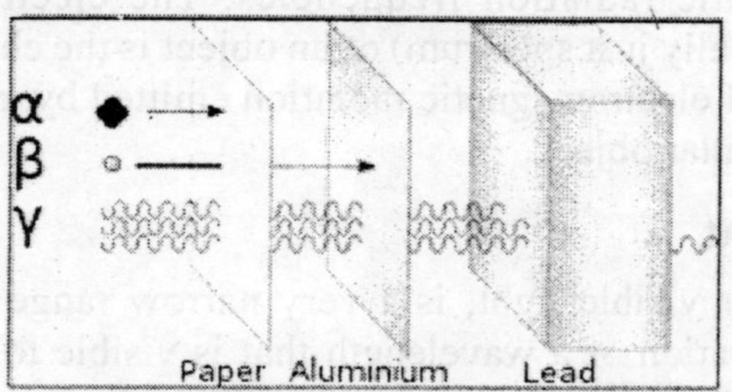

Fig. 1. Penetration Power of Three Ionising Radiations

Neutron Radiation

Neutron radiation is sometimes called "indirectly ionizing radiation" since so many of its interactions with matter eventually result in ionization. Neutron radiation consists of free neutrons. These neutrons may be emitted during either spontaneous or induced nuclear fission, nuclear fusion processes, or from any other nuclear reactions. It does not ionize atoms because neutrons have no charge. However, both slow and fast neutrons react with the atomic nuclei of many elements and therefore inducing radioactivity. This process is known as neutron activation.

Electromagnetic Radiation

Electromagnetic radiation (sometimes abbreviated EMR) takes the form of self-propagating waves in a vacuum or in matter. EM radiation has an electric and magnetic field component which oscillate in phase perpendicular to each other and to the direction of energy propagation. Electromagnetic radiation is classified into types according to the frequency of the wave, these types include (in order of increasing frequency): radio waves, microwaves, terahertz radiation, infrared radiation, visible light, ultraviolet radiation, X-rays and gamma rays.

The ability of an electromagnetic wave (photons) to ionize an atom or molecule depends on its frequency. Radiation on the short-wavelength end of the electromagnetic spectrum—high frequency ultraviolet, X-rays, and gamma rays—is ionizing.

The electromagnetic spectrum is the range of all possible electromagnetic radiation frequencies. The electromagnetic spectrum (usually just spectrum) of an object is the characteristic distribution of electromagnetic radiation emitted by, or absorbed by, that particular object.

Visible Light

Light, or visible light, is a very narrow range of electromagnetic radiation of a wavelength that is visible to the human eye (about 400–700 nm), or up to 380–750 nm.

Infrared

Infrared (IR) light is electromagnetic radiation with a wavelength between 0.7 and 300 micrometres. IR wavelengths are longer than that of visible light, but shorter than that of terahertz radiation microwaves. Bright sunlight provides an irradiance of just over 1 kilowatt per square meter at sea level. Of this energy, only 527 watts is infrared radiation and he rest is other types of radiation.

Microwave

Microwaves are electromagnetic waves with wavelengths ranging from 300 MHz (0.3 GHz) and 300 GHz. This includes both UHF and EHF (millimeter waves).

Radio Waves

Radio waves are a type of electromagnetic radiation with wavelengths in the electromagnetic spectrum longer than infrared light. Like all other electromagnetic waves, they travel at the speed of light. Naturally-occurring radio waves are made by lightning, or by astronomical objects. Artificially-generated radio waves are used for fixed and mobile radio communication, broadcasting, radar and other navigation systems, satellite communication, computer networks and innumerable other applications.

Very Low Frequency (VLF)

Very low frequency or VLF refers to radio frequencies (RF) in the range of 3 to 30 kHz. Since there is not much bandwidth in this band of the radio spectrum, only the very simplest signals are used, such as for radio navigation.

Extremely Low Frequency (ELF)

Extremely low frequency (ELF) is a term used to describe radiation frequencies from 3 to 30 Hz. In atmosphere science, an alternative definition is usually given, from 3 Hz to 3 kHz.

Thermal Radiation (heat)

Thermal radiation, a common synonym for infra-red, is the process by which the surface of an object radiates its thermal energy in the form of electromagnetic waves. Infrared radiation from a common household radiator or electric heater is an example of thermal radiation, as is the heat and light (IR and visible EM waves) emitted by a glowing incandescent light bulb. Thermal radiation is generated when heat from the movement of charged particles within atoms is converted to electromagnetic radiation. The emitted wave frequency of the thermal radiation is a probability distribution depending only on temperature, and for a black body is given by Planck's law of radiation.

Black Body Radiation

Black body radiation is radiation from an idealized radiator that emits at any temperature the maximum possible amount of radiation at any given wavelength. A black body will also absorb the maximum possible incident radiation at any given wavelength. The radiation emitted covers the entire electromagnetic spectrum and the intensity (power/unit-area) at a given frequency is dictated by Planck's law of radiation.

Uses of Radiation

In Medicine

Radiation and radioactive substances are used for diagnosis, treatment, and research. X-rays, for example, enable doctors to find broken bones and to locate cancers that might be growing in the body as shown. Doctors also find certain diseases by injecting a radioactive substance and monitoring the radiation given off as the substance moves through the body as shown in Fig. 2. Ionizing Radiations are used for various types cancer treatment and different types of therapy.

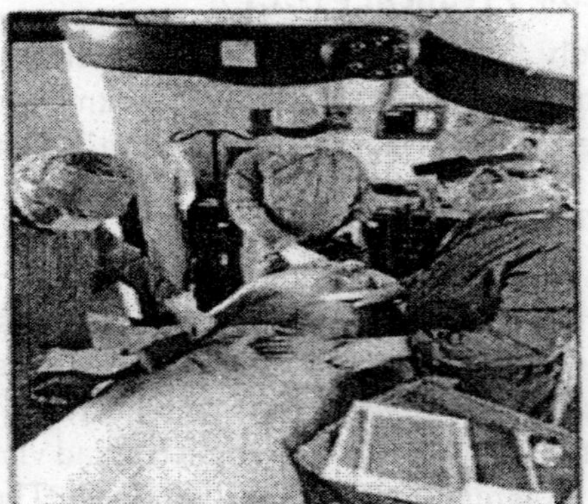

Fig. 2 Use of Radiation for Treatment

In Communication

All modern communication systems use forms of electromagnetic radiation. Variations in the intensity of the radiation represent changes in the sound, pictures, or other information being transmitted. For example, a human voice can be sent as a radio wave or microwave by making the wave vary to correspond variations in the voice. A typical antenna for receiving signals is shown in Fig. 3.

Fig. 3 Use of E.M. Radiation for Communication

In Science

Researchers use radioactive atoms to

determine the age of materials that were once part of a living organism. The age of such materials can be estimated by measuring the amount of radioactive carbon they contain in a process called radiocarbon dating. Environmental scientists use radioactive atoms known as tracer atoms to identify the pathways taken by pollutants through the environment.

Radiation is used to determine the composition of materials in a process called neutron activation analysis. In this process, scientists bombard a sample of a substance with particles called neutrons. Some of the atoms in the sample absorb neutrons and become radioactive. The scientists can identify the elements in the sample by studying the emitted radiation.

Radiation Hazards

In the above sections we have discussed about the different types of radiations and their uses. Thus we come to know how useful they are but at the same time, they do have some adverse effects also. Here we present some causes of radiation exposure/ contamination and their effects.

Causes of Radiation Exposure/Contamination

Nuclear and Radiation Accident

A nuclear and radiation accident is defined by the International Atomic Energy Agency as "an event that has led to significant consequences to people, the environment or the facility. Example of a "major nuclear accident" is one in which a reactor core is damaged and large amounts of radiation are released, such as in the Chernobyl Disaster in 1986 & recently occurred Japan (Fukushima) Disaster (Fig. 4) in 11 March 2011.

Fig. 4 Fukushima Nuclear Plant

Benjamin K. Sovacool has reported that worldwide there have been 100 accidents at nuclear power plants from 1952 to 2011. There have been comparatively few fatalities associated with nuclear power plant accidents.

Loss of proper disposal of medical or industrial radioactive sources may lead to serious injuries and harm to the people as well as their surroundings. For example the recently occurred incident in west Delhi industrial area(Mayapuri) in which radiations of Cobalt-60 injured five persons including a scrap dealer.

Terrorist Event

- Radiological dispersal device (dirty bomb)
- Low-yield nuclear weapon (Fig. 5)
- Attack on or sabotage of a nuclear facility (Fig. 6)

Fig. 5 Low-yield nuclear weapon

Fig. 6 Sabotage of a nuclear facility

Effect of Radiation on Materials and Devices

Radiation may affect materials and devices in deleterious ways:

- By causing the materials to become radioactive and thus having the potential to cause radiation poisoning.
- By nuclear transmutation of the elements within the material.

- By radiolysis within the material, which can weaken it, cause it to swell, polymerize, promote corrosion, cause belittlements, promote cracking or otherwise change its desirable mechanical, optical, or electronic properties.
- By formation of reactive compounds, affecting other materials.
- By ionization, causing electrical breakdown, particularly in semiconductors employed in electronic equipment, with subsequent currents introducing operation errors or even permanently damaging the devices.

Effect on Gases

Exposure to radiation causes chemical changes in gases. The least susceptible to damage are noble gases, where the major concern is the nuclear transmutation with follow up chemical reactions of the nuclear reaction products. High-intensity ionizing radiation in air can produce a visible ionized air glow of telltale bluish-purplish color. The glow can be observed e.g. during criticality accidents.

Effect on Liquids

Like gases, liquids lack fixed internal structure; the effects of radiation is therefore mainly limited to radiolysis, altering the chemical composition of the liquids. All liquids are subject to radiation damage, with few exotic exceptions; e.g. molten sodium, where there are no chemical bonds to be disrupted, and liquid hydrogen fluoride, which produces gaseous hydrogen and fluorine, which spontaneously react back to hydrogen fluoride.

Effect on Water

Water subjected to ionizing radiation forms free radicals of hydrogen and hydroxyl, which can recombine to form gaseous hydrogen, oxygen, hydrogen peroxide, hydroxyl radicals, and peroxide radicals. In living organisms, which are composed mostly of water, majority of the damage is caused by the reactive oxygen species, free radicals produced from water. The free radicals attack the biomolecules forming structures within the cells, causing oxidative stress (a cumulative damage which may be significant enough to cause the cell death, or may cause DNA damage possibly leading to cancer).

Effect on Humans

All types of radiation produces changes in the living tissues. The resultant cellular injuries causes physiological and pathological changes leading to "Radiation sickness". The radiation effects may be somatic or genetic. Somatic effects are harmful in his lifetime whereas genetic effect affects generations.

Radiation may causes changes in complex molecular systems of living cells, primarily in the following two ways-

- **Direct interactions :** These appear when the molecules of body are modified due to absorption of the energy of the incident photons of X-rays or any other radiations.
- **Indirect interactions :** These appear due to the damage caused by the product of radiation decomposition (Radiolysis)of water and other solutes of the body.

Direct Interactions: These are brought about by following three important mechanisms-

Thermalization

This is a process where incident radiations (X-ray photons) agitate the living macromolecules in a non-destructive manner and in the process transfer their kinetic energy and there-by increase the temperature of the molecular system.

Excitation

This is a process whereby the incident radiation while traversing through the tissue force bound electrons to be "knocked" free from their parent atoms or molecules. These free electrons are then available to interact with other atoms and molecules within the irradiated system. Generally however the free electrons are quickly captured and their energy added to the overall thermal energy of the system with little or no other effect.

Ionization

This is a process whereby molecules absorb sufficient energy from radiation to break their molecular bonds. This causes the direct modification or destruction of complex molecules.

- Covalent bonds and ionic bonds, the bonds between the atoms that form simple inorganic molecules such as sodium chloride (common table salt) have binding energies of about 2 to 5eV, about one half the strength of the ionizing energies of individual atoms.
- Organic molecules, many of which are important in cell biology, have bond energies of about 0.04 to 0.3eV and are typically referred to as "weak bonds".

***Indirect interactions**:* The most abundant substances cell is water. It is because of this ubiquitous nature of the body, that the most of the energy of incident radiation is absorbed by the water molecules. These water molecules may sometimes be broken into highly reactive components particularly when high energy radiation is absorbed. The reactive components of the modified molecule are then available to react with complex organic molecules within the cell and causes damage.

So care should be taken while coming under the influence of radiation. If by mistake one has absorbed radiation the following symptoms are seen:

Symptoms of Radiation Poisoning

- Bleeding from the nose, mouth, gums and rectum
- Bloody stool
- Bruising
- Dehydration
- Diarrhea
- Fainting
- Fatigue
- Hair loss
- Mouth ulcers

- Weakness
- Inflammation of exposed areas (redness, tenderness, swelling, bleeding)
- Nausea and vomiting
- Open sores on the skin
- Skin burns (redness, blistering)
- Sloughing of skin
- Ulcers in the esophagus, stomach or intestines

Severity of symptoms depends on exposure duration and strength. Below in table 1, we present radiation measurements of various events.

Remedies and Available Safety Measures

In the above section we have seen the causes and effects of radiations. Thus we come to know that radiation can cause many harmful effects and in the meantime it could be proved useful if it is used under safety limits. So following are safety measures which should be taken to avoid such accidents:

- Disposal of medical or industrial radioactive sources and wastes should be proper.
- Cell phones should not be kept near one's heart or near head while sleeping.

Remedies while Dealing With Radiations at Special Work Practices

X-Rays

General Principles of radiation safety-

Increase the distance between radiation source and personnel

- Only the minimum required personnel should be present inside X-ray room during exposure.
- The patient should preferably be restrained by chemical methods and not manually.

- A cassette holding device should be used whenever possible.
- The operator should be behind shielding screen or at least 6 feet away from X-ray source.

Use of protective barriers

- Protective barriers are necessary to protect oneself from scattered radiation.
- Aprons – Aprons should have a minimum of 0.25 mm lead equivalent for voltage up to 100 kV.
- Gloves and goggles – Lead equivalent should not be less than 0.33 mm for voltage up to 100 kV.
- X-ray room and shield equipments-
  - The wall of X-ray room should be of at least 22 cms thick concrete.
  - The wall should be painted with radiation absorbent paint.
  - When there is possibility of X-ray beam consistently directed horizontally, the wall should have a lead lining sandwiched between plywood.
  - Whenever possible X-ray operator should be in a lead shielded cabin during exposure.

Use of optimal exposure factors and reduction of unnecessary radiography

- Different radiographic accessories which reduce exposure factors (e.g. intensifying screens) should be used along with correct exposure factors.
- Radiography should not be done just because of owner's wish; a radiologist should decide its necessity.

Use of radiation monitoring devices

- The personnel involved in regular radiography should wear a radiographic monitoring device (Fig. 7) so that total exposure to that person can be ascertained and accordingly he/she may be rested from radiological work in case of overexposure than the prescribed limits.

Events	*Radiation Reading, Millisievert (mSv)*
Accumulated dosage estimated to cause a fatal cancer many years later in 5% of people	1,000.00
Airline crew flying New York to tokyo polar route, annual exposure	9.00
CT scan: abdomen & pelvis	15.00
CT scan: head	2.00
CT scan: heart	16.00
Chest x-ray	0.10
Dental x-ray	0.01
Dose in full-body CT scan	10.00
Exposure of Chernobyl residents who were relocated after the blast in 1986	350.00
Lowest annual dose at which any increase in cancer is clearly evident	100.00
Mammogram breast x-ray	0.40
Max radiation levels recorded at Fukushima plant yesterday, per hour	400.00
Natural radiation we're all exposed to, per year	2.00
Radiation per hour detected at Fukushimia site, 12 March	1.02
Recommended limit for radiation workers every five years	100.00
Single does which would kill half of those exposed to it within a month	5,000.00
Single dosage which would cause radiation sickness, including nausea, lower white blood cell count. Not fatal	1,000.00
Single dose, fatal within weeks	10,000.00
Spine x-ray	1.50
Typical dosage recorded in those Chernobyl workers who died within a month	6,000.00

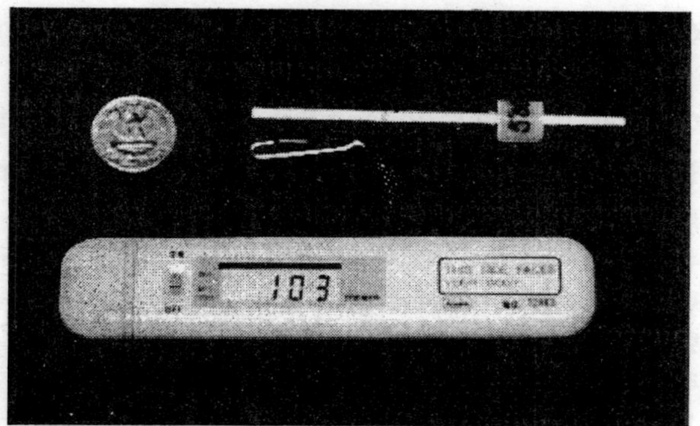

Fig. 7 A Radiographic Monitoring Device

- A busy radiographer should get his badg checked every month; a less busy radiographer may get it checked every three month.

Consideration for sex and age of the personnel involved

- Person under 18 years of age should not be involved in radiographic work because of the sensitivity of their growing tissues.
- Similarly due to sensitivity of ova and embryos at certain stage of development, pregnant, potentially pregnant and women during menstrual cycle avoid X-ray work.

UV-Rays

Never allow the skin or eyes to be exposed to UV radiation sources. The UV radiation generated by laboratory equipment can exceed recommended exposure limits and cause injury with exposures as brief as three seconds in duration.

Biological Safety Cabinets – Never work in a biological safety cabinet while the germicidal lamp is on. If possible, close the sash while lamp is on.

Transilluminators – Never use a transilluminators without the protective shield in place. Shields must be kept clean and replaced when damaged.

Crosslinkers – Crosslinkers must not be used if the door safety interlock is not working properly.

One can use the following things for radiation safety:

- **Protective Clothing**: Wear standard laboratory apparel including a fully buttoned lab coat, long pants and closed toe shoes. While working with UV radiation sources, lab workers must be particularly vigilant to prevent gaps in protective clothing that commonly occur around the neck and wrist areas.
- **Eye/Face Protection**: If there is any potential for the eyes and face to be exposed to UV radiation, a polycarbonate face shield stamped with the ANSI Z87.1.-1989 UV certification must be worn to protect the eyes and face. Ordinary prescription eye glasses may not block UV radiation. UV certified goggles and safety glasses will protect the eyes, but it is common for lab workers to suffer facial burns in the areas not covered by the goggles or glasses.
- **Gloves**: Wear disposable nitrile gloves to protect exposed skin on the hands. Ensure wrists and forearms are covered between the tops of gloves and the bottom of the lab coat sleeves.

Conclusions

Natural sources of radioactivity are all around, and man-made radioactive materials are a vital part of medicine and industry. Exposure to some radiation, natural or man-made, is inevitable. We live with radiation everyday, therefore we must understand both its risks and benefits.

These radiations move with such great velocities as that of light, so it engulfs all the geographical, demographic scapes of the sphere as well as the territorial waters of all the countries (all oceans inclusive). Every sphere is affected by the radiation.

In this chapter, we have tried to throw some light on the various radiation hazards that took place at various places across the globe. This study enables us to compare the different radiation hazards and get a picture of how harmful these radiations are and what levels of destruction has been caused by them. The study includes both the causes and effects of radiation hazards along with the remedies and safety measures available that are associated with it.

References

1. http://en.wikipedia.org/wiki/Radiation.
2. http://www.epa.gov/rpdweb00/understand/health_effects.html.
3. http://en.wikipedia.org/wiki/Radiation_damage.
4. **Tyagi , Dr. S.P.,** *Radiation Hazards and Safety.*
5. Health Physics Society Public Education Website.
6. **Prof. Richard Muller,** *Physics for Future Presidents*, Webcast. Berkeley
7. Nuclear Disasters - Slideshow by Life Magazine.
8. International Atomic Energy Agency website with extensive online library
9. **Dr. Matthias Braun,** *The Fukushima Daiichi Incident*, March 31, 2011 © AREVA 2011.
10. **Kereiakes JG, Rosenstein M.,** *Handbook of Radiation doses in Nuclear Medicine and Diagnostic x-ray.* Boca Raton, FL: CRC Press; 212-213; 1980.
11. National Research Council, Health Effects of Exposure to Low Levels of Ionizing Radiation (BEIR V). Washington DC: National Academies Press; 1990.
12. http://www.brown.edu/Administration/EHS/laser.
13. website: www.epa.gov/radiation
14. "Radiation Injury", Encyclopedia Britannica 2011 Ready Reference CD. Chicago: Encyclopedia Britannica, 2011.